# Chemistry 151

# Laboratory Exercises

**Fundamental Experiments in Chemistry**
**Fifth Edition**

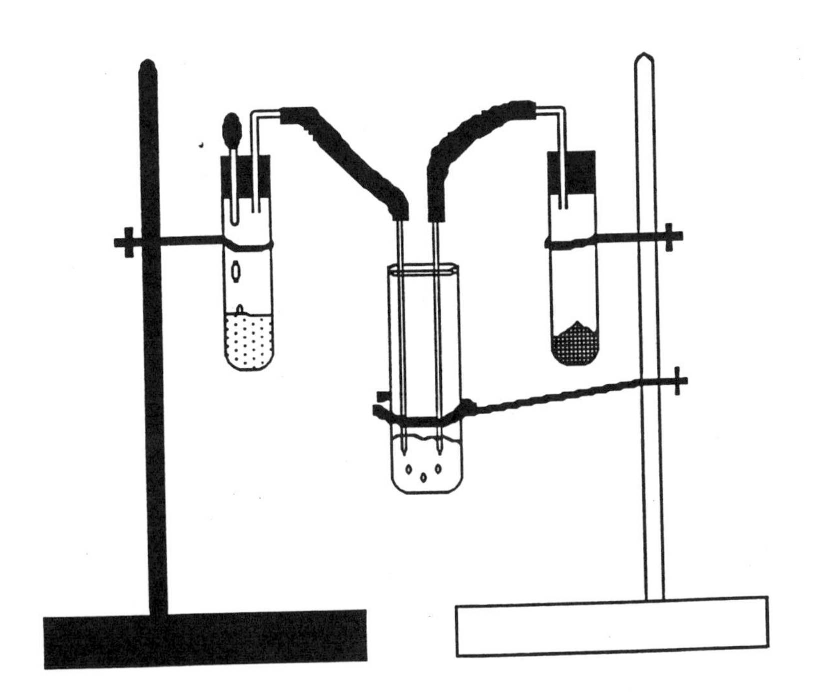

**Karen E. Eichstadt**
**Ohio University**

KENDALL/HUNT PUBLISHING COMPANY
4050 Westmark Drive Dubuque, Iowa 52002

This edition has been printed directly from camera-ready copy.



ISBN 0-7872-2746-3



Printed in the United States of America
10 9 8 7 6 5 4 3 2 1

# Table of Contents

Disclaimer ---------- iv

To the Student---------- v

Laboratory Agreement---------- vi

Laboratory Equipment---------- vii

Experiments

1. Density---------- 1
2. Separation of a Mixture ---------- 9
3. A Study of Hydrates ---------- 17
4. Chemical Equations---------- 25
5. Reactions in Solution Studied by Microscale---------- 35
6. Redox Reactions---------- 47
7. Calorimetry ---------- 57
8. A Redox Titration ---------- 67
9. Determination of an Empirical Formula ---------- 75
10. The Hydrogen Spectrum ---------- 83
11. A Chemical Family, the Halogens ---------- 87

Appendices ---------- 95

## Disclaimer

These experiments have been written within the guidelines of sources believed to be reliable and within the safety standards of common laboratory practice. The safety information does not purport to specify minimum legal standards. Additional information regarding safe practices must be added at the prelaboratory lecture which all students are required to attend before entering the laboratory.

# To the Student

You are about to begin your study of college chemistry. The laboratory will enable you to investigate at close-hand principles discussed in the lecture. By making observations and generalizing, your understanding of chemistry will be enhanced. You will also meet your Teaching Assistant (TA) who is a graduate or advanced undergraduate student. The TA will have regular office hours to assist you. Take advantage of this opportunity for personal help.

Our standard laboratory has the following components:

a) *the prelaboratory assignment:* a written assignment located on the last pages of the experiment to be completed before coming to lab. After reading the procedure, you should be able to answer the questions on the pre-lab page. You will turn it in at the beginning of lab.

b) *prelaboratory lecture:* a brief discussion of the day's exercise with special emphasis on safety and techniques. You will meet in a large room as indicated on your schedule at the beginning of every laboratory period. After the pre-lab lecture (15-20 minutes), we will proceed to the ground floor laboratories in Clippinger.

When you arrive at the pre-lab lecture you should have the pre-laboratory assignment completed and ready to turn in. No one will be admitted to the laboratory without having attended the pre-lab lecture.

c) *laboratory experiment:* the experiment to be performed as directed in the laboratory manual. If you have questions regarding procedures, ask your TA. Record all data carefully. Make systematic calculations.

d) *post-laboratory questions:* wrap-up questions at the end of each experiment. The complete lab report must be turned in before leaving the laboratory.

You will be graded on general technique, the answers to questions on the pre- and post-laboratory assignments and your experimental results. Some experiment have "unknowns" which you are to identify. Take care in your analysis because your results are a large segment of your grade.

Most students find their laboratory experience enjoyable. Get to know your fellow students. You might form an informal study group.


## Acknowledgements:

I gratefully acknowledge the suggestions from students and TA's in evaluating the earlier editions of this manual. I welcome your comments which will provide insights into methods and techniques that can improve General Chemistry lab.

K.E.E.

## Chemistry Laboratory Agreement

Observance of the rules is necessary for a safe, clean laboratory environment for all students. Failure to comply with the *safety rules* below will result in expulsion from the laboratory. Penalties for violation of *housekeeping rules* range from deduction of points to expulsion from the course.

**Safety Rules:**

1. Wear safety goggles at ALL times.
2. Know the exact location and operation of all safety equipment. Use the equipment only when necessary. Tampering with safety equipment at any time violates Student Conduct Code A.13. (O.U. Student Handbook.)
3. Report all injuries immediately to the laboratory instructor, no matter how small.
4. Never work unsupervised in the laboratory.
5. Enter the laboratory properly dressed/groomed for maximum protection. No bare feet , thongs, or shoes with exposed toes. Keep long hair/beards away from flames.
6. Never eat, drink, smoke, or chew in the laboratory.
7. Do not perform any unauthorized experiments.

**Housekeeping Rules**

1. Place belongings such as bookbags, coats, etc in the designated areas.
2. Keep the laboratory clean at all times. Clean up any spilled chemical immediately according to procedures in the experiment . Sweep up and replace any broken glassware. Inform TA immediately of spillage of hazardous material, broken thermometers, or other special glassware.
3. Put ALL equipment in its proper place at the end of the laboratory period.
4. Do not tamper with laboratory equipment. Use the balances, centrifuges, and other permanent pieces of equipment in the proper manner. Report any malfunction to the TA. See Student Conduct Code A, OU Student Handbook.
5. Do not take reagent bottles to your work area. Use smaller containers to obtain the proper amount from the supply shelf. Do not contaminate the supply containers. Do not return any unused chemical to the bottle.
6. Chemicals must be disposed of as directed in the laboratory experiment. Compounds that are non-hazardous and soluble may be washed down the sink with excess water. Compounds which are hazardous must be disposed in the proper containers.
7. Dispose of other waste materials properly. Used matches, filter paper, litmus paper, etc may go in the crocks. Broken glass must be placed in the "broken glass" container.

**Storeroom Operation and Breakage Card Rules:**

1. The student is responsible for all equipment issued at check-in.
2. All students are required to have a breakage card. Any item broken or missing must be replaced immediately by presenting a breakage card at the storeroom window. No cash is accepted. Breakage cards are purchased at the Cashier's Office ($20) in Chubb Hall.
3. A $6 consumable supplies charge will be assessed at mid-quarter by punch out of the breakage card. Any unused portion of the breakage card may be redeemed at the completion of the course or carried on to the next course.
4. Students who do not properly check-out will not be allowed to receive a grade in the course.

**Keep this copy with you for all laboratory exercises.**

# Laboratory Equipment

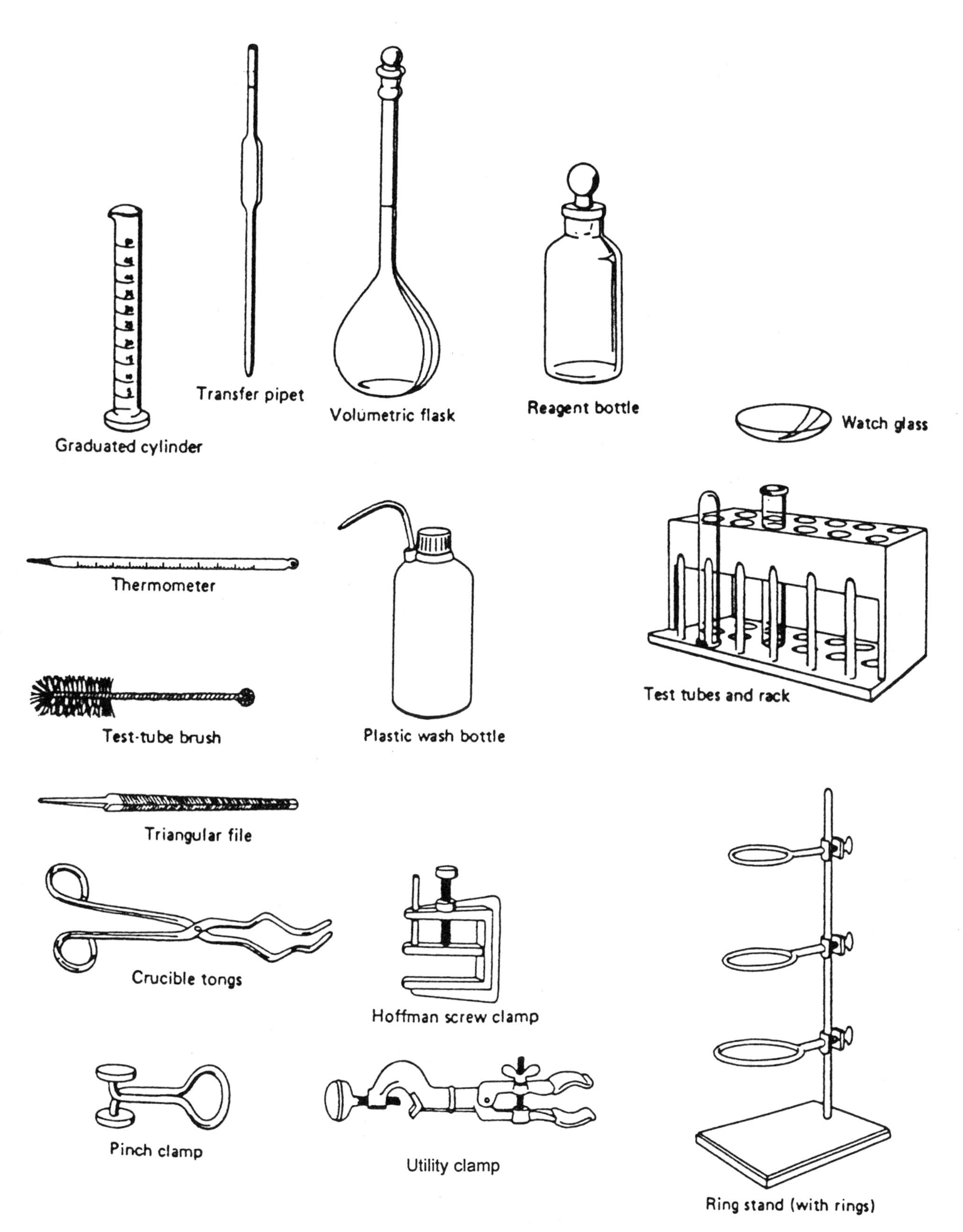

# Laboratory Equipment

Buret

Stainless steel scoop

Evaporating dish

Triangle

Test-tube holder

Crucible with cover

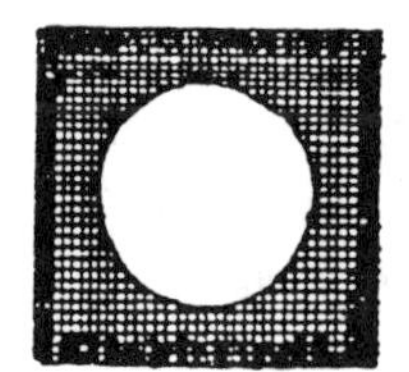

Wire gauze (flame resistant center)

# Experiment 1
# Density of Liquids and Solids

Density is a physical property of substances, the ratio of mass to volume.

$$\text{density} = \frac{\text{mass}}{\text{volume}} \qquad D = \frac{m}{V}$$

If a scientist knows two of these quantities, the third may be calculated from the equation. Sometimes, density is the quantity desired. In that case, precise mass and volume measurements are made and the density calculated. At other times, a chemist may know the density of a substance being used. If the volume is measured directly, then the mass can be calculated. This procedure is particularly useful for liquids where the density can be looked up in a handbook. It is usually more convenient to measure the volume of a liquid than to measure its mass.

**About weighing:** Our laboratories are equipped with top-loading balances that have a digital readout. You may measure a *thousandth of a gram* so typical readings are 44.854 g, 85.200 g etc. Even with a digital balance you will notice that there is uncertainty in the last digit just as in any other measurement. Record what seems to be the best digit and realize in calculations that there is uncertainty in the value. *The uncertain digit is significant.*

Use the same balance for all weighings. Check to see that it is level but do NOT adjust it. Ask the TA if help is needed.

DO NOT weigh directly on the pan. Always use a container or a piece of weighing paper.

Most of our balances have a *tare* feature. This allows you to "subtract" the mass of the container you are using automatically. The feature is useful if you are merely going to weigh out an amount of a substance but will not be returning to the balance for future weighings.

Example: "Weigh a 2.000 g sample of NaCl"

You would put a piece of weighing paper on the balance, tare it by pushing the tare lever, then sprinkle the solid on the paper.

But do NOT use the tare if you are going to make several weighings with the same container.

Example: "Weigh the empty crucible. Add sample. Heat. Reweigh...."

In this case several readings with the crucible will be needed, do not use the tare.

The volume will be measured in several ways. In Part A, the volume of a liquid will be measured directly in a graduated cylinder. In Part B, the volume of a solid will be measured by displacement of water in a graduated cylinder. In Part C, a volumetric pipet will be used. **The pipet is designed to deliver a precise volume with two decimal places, such as 5.00 mL and 10.00 mL.**

**About measuring liquids:** Liquids form a curved surface called a *meniscus* within a container. The smaller the diameter of the container, the more noticeable the curved surface is . Read the bottom of the meniscus while observing at eye level.

## Use of the Pipet

The pipet is an instrument that will accurately deliver a specified volume. The term "TD" painted on the pipet means "to deliver."

Caution: Do **not** use the mouth to draw solutions into the pipet. Always use a bulb.

1. Squeeze the air out of the bulb.
2. Place the bulb over the top of the pipet.
3. Draw liquid into the pipet to a level above the calibration mark.
4. Remove the bulb and quickly put your finger on the top of the pipet.
5. Gently release pressure with the finger until the bottom of the meniscus (the curved surface) is on the calibration line.
6. Take your finger off the pipet and allow the liquid to drain out. Touch the pipet to the side of the vessel to remove the liquid at the tip. There will be a small amount of liquid remaining in the pipet. Leave it. Do not blow it out. The pipet is calibrated "to deliver" exactly the specified amount.

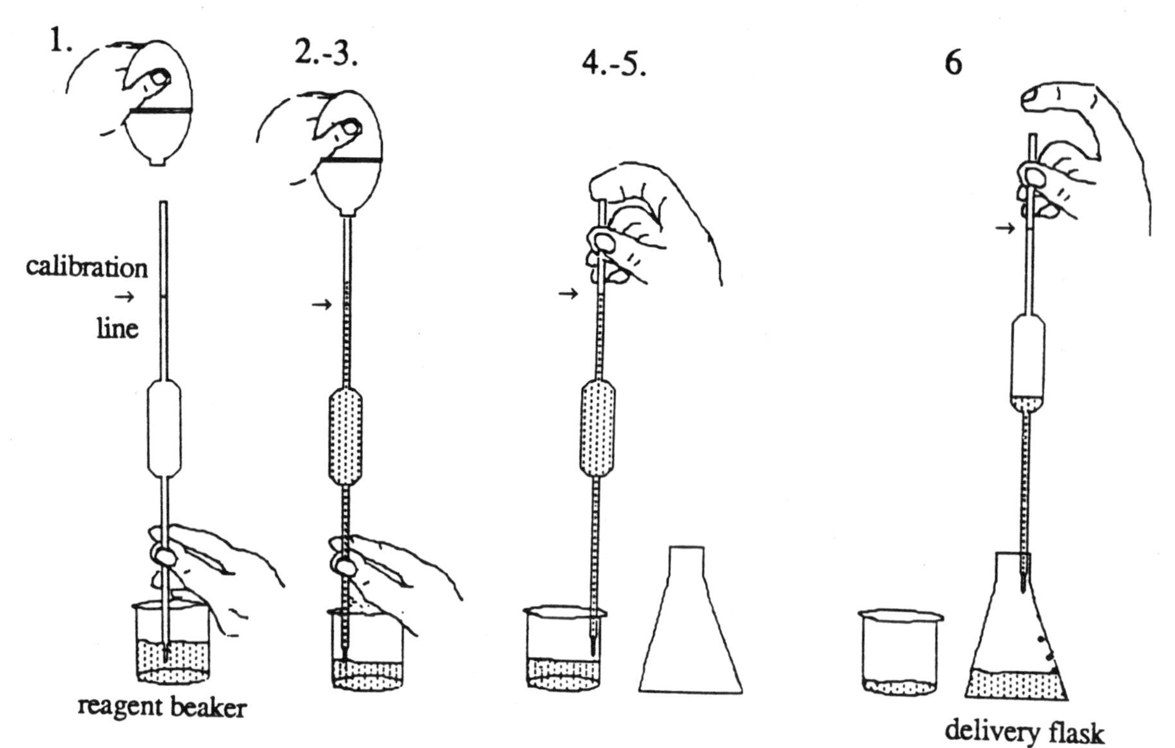


Figure 1: Technique for Using a Pipet

## Significant Figures in Observations and in Calculations

Significant figures are used by a scientist to be report exactly what was **measured**, no more, no less. The number recorded must reflect the measuring tool, such as the ruler used, the balance, the thermometer, etc. When calculations are performed with measured values, the operational rules are also applied.

Exact numbers, from a definition or from counting, are *not measured* with a tool and have an unlimited number of significant figures. Examples of exact numbers are 1 ft = 12 inches, 7 days = 1 week.

| Rule | Example | Number of Sig Figs |
|---|---|---|
| 1. All **measured** digits are significant. | 3.24 cm | three |
| 2. Zeros may or may not be significant. a) zeros between other digits are significant | 3.004 grams | four |
| b) zeros that are measured are significant | 43.00 cm | four |
| | 1.200 grams | four |
| | $6.02 \times 10^{23}$ atoms | three |
| c) zeros that are merely placeholders are NOT significant | 0.00789 km | three |
| | 0.00456 km = 4.56 m | three |
| | 2.7000 kg | five |
| | (crude estimate) 10,000 ℓ | one |
| 3. Use scientific notation to indicate significant figures clearly | $5.00 \times 10^2$ | three |
| | $5.000 \times 10^2$ | four |
| 4. In MULTIPLICATION and DIVISION the answer may have as many *significant figures* as the least known factor. | 54.222 m / 0.508 sec = 107 m/sec (5 sig figs) (3 sig figs) | three |
| 5. In ADDITION and SUBTRACTION the answer may have as many *decimal places* as the least known addend. | 14.246 g + 86.854 g = 101.100 g | six* |
| *three decimal places required so six sig figs even though both addends had 5 sig figs | | |
| 6. In combined operations, both rules #4 and #5 apply | (24.987 g - 24.087 g) / 10.00 mL = $9.00 \times 10^{-2}$ g/mL | |
| (24.987 g - 24.087 g) = 0.900 g (3 sig figs) | 10.00 mL ( 4 sig figs) | |

## Procedure:

Part A. Determination of the Density of an Unknown Solid.

Obtain a metal sample from the bottle. These are non-homogeneous brass samples. Each has a different density. Be careful not to handle it with you hands. Use tongs and a towel or Kim-wipes. Record the number of your sample on your data page.

1. Weigh the sample.

2. Place 50.0 mL water in your graduated cylinder. Insert the brass sample into the water carefully by tipping the cylinder and allowing the sample to slide down the side. (Don't drop it in because it might break the bottom of the cylinder). Be careful not to have any air trapped underneath the sample. Record the new volume of the liquid level. Calculate the volume of the brass sample by water displacement. Note the **allowable significant figures.** Calculate the density of the metal sample.

3. Repeat the procedure in section one after carefully drying the metal sample.

4. Average the density determinations from #2 and #3 above.

Part B: Determination of the Density of an Unknown Salt (NaCl) Solution.

In this analysis the volume will be determined more accurately than was possible in Part A. A volumetric pipet will be used according to the procedure outlined on page 2 with a bulb and the index finger. **Practice with water until you have the technique mastered.** Then continue with your unknown solution.

Your TA will assign an unknown salt solution to you (151-Z, 151-B etc). Fill a 100 mL beaker about half full of the unknown solution that you will be using. You need enough so that you can pipet without getting air bubbles into the pipet.

1. Weigh a clean dry 50 mL beaker accurately.

2. With the 5.00 mL pipet deliver 5.00 mL into the weighed container. Reweigh. Calculate the density. Note the **allowable significant figures.** Repeat the determination and report the average density after carefully drying the container.

3. Using the 10.00 mL pipet, deliver 10.00 mL into a preweighed container. Reweigh. Calculate the density. Note the **allowable significant figures.** Repeat the determination and report the average density after carefully drying the container.

# Experiment 2
# Separation of a Mixture

## Introduction:

The separation of mixtures is a frequent problem for a chemist. By careful consideration of the properties of the substances, the scientist can design a suitable plan of purification.

In this experiment the separation of a mixture of sand ($SiO_2$), salt (NaCl), and ammonium chloride ($NH_4Cl$) will be attempted. A difference in physical properties will allow separation of the components of the mixture and determination of the % composition.

Some data from *The Handbook of Chemistry and Physics* is listed in Table 1. We see that NaCl and $NH_4Cl$ are both soluble in water but NaCl has a high melting point and $NH_4Cl$ doesn't melt at all but it sublimes! This property may be the key to the separation of these two substances. With heating $NH_4Cl$ does not melt, but rather goes directly into the vapor state at about 340°C. That temperature is well below the melting point of NaCl so we should be able to get good separation.

Table 1:

| Compound | Solubility (g /100g $H_2O$ at 25°C) | Melting Point °C | Appearance |
|---|---|---|---|
| NaCl | 36 | 801 | white crystals |
| $NH_4Cl$ | 37 | subl 340 | white crystals |
| $SiO_2$ | 0 | >1600 | white crystals |

A feasible plan is diagrammed in Figure 1. First the $NH_4Cl$ is sublimed. The solid residue is a mixture of the two compounds, NaCl and $SiO_2$. Water is then added to dissolve the NaCl and any small amount of $NH_4Cl$ that may have been retained. After drying pure $SiO_2$ can be weighed. The differences in mass at each step will allow us to calculate the mass of each component of the mixture.

The apparatus needed to sublime the $NH_4Cl$ is pictured. The vapors will condense in the funnel as a "white cloud" and be carried off to the water system where the $NH_4Cl$ dissolves.

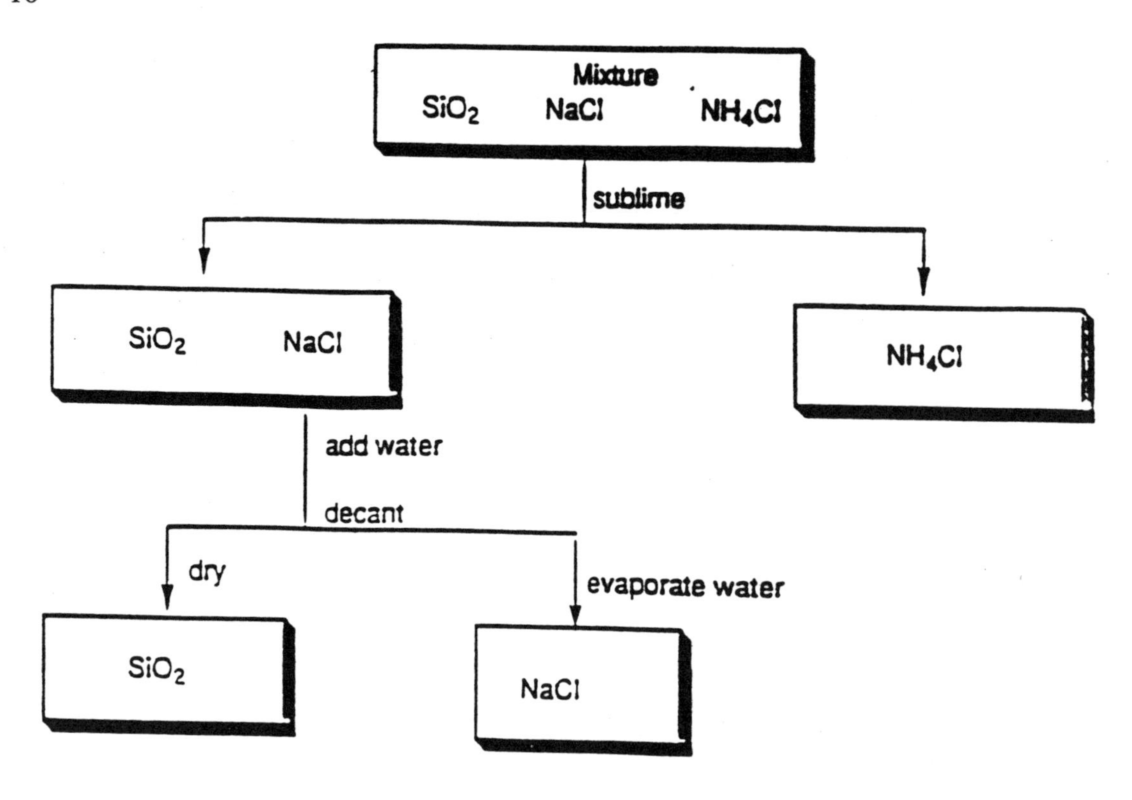


Figure 1: Scheme for the Separation of a Mixture

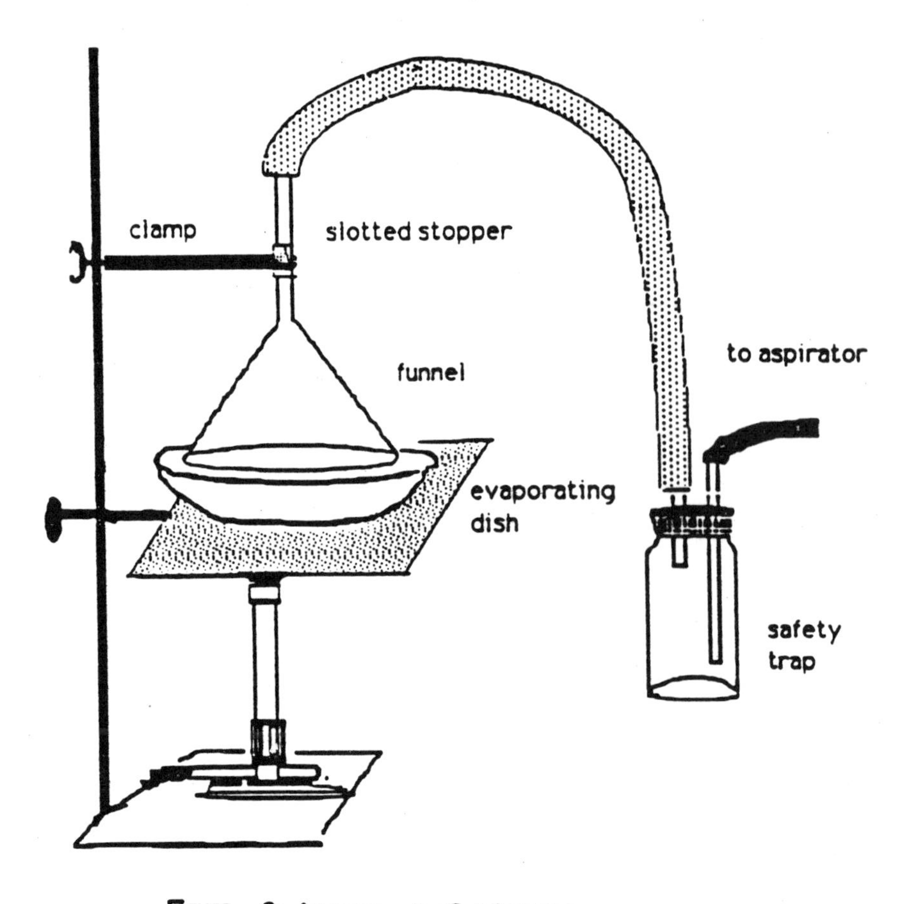


Figure 2: Apparatus for Sublimation

**Procedure:** **Work in pairs. Each member will do an analytical weighing. One in Step 7. The other in Step 9.**

1. Obtain an unknown sample. Record its number. Shake it vigorously before weighing.

2. Weigh an evaporating dish to the nearest 0.001 g. Place the dish on a wire gauze and handle it with tongs so no moisture from your hands gets on it.

3. Weigh accurately ALL of a sample of the unknown assigned to your group. You need to know the weight accurately (9.556 g, 11.675 g are typical masses within the guidelines.)

4. Place the evaporating dish and sample on a wire gauze as shown in Figure 2. Invert a funnel over the dish to catch the fumes of $NH_4Cl$ as it sublimes. The funnel is attached with a rubber hose to a trap that is attached to an aspirator. Heat the dish slowly at first. You will see white fumes in the funnel which then dissolve in the water in the sink. Heat more strongly until no more fumes are observed. This procedure takes about 15 minutes. Remove the funnel. Heat for another 2-3 minutes.

5. Cool the dish. Weigh. This is the mass of the NaCl-$SiO_2$ mixture. The difference between the first weighing and the second is the mass of the $NH_4Cl$ that was removed.

6. Add 10 mL of distilled water to the dish and swirl. Decant the water into a beaker being careful not to lose any of the $SiO_2$. Again add 10 mL of clean distilled water and repeat. Two more washings (about 3 mL each) are necessary in order to be sure that only $SiO_2$ remains. **Incomplete washings are a major source of error in this experiment.** SAVE ALL THE WASHINGS (about 26 mL total) in the beaker. See Step 9.

7. Heat the remaining $SiO_2$ in the evaporating dish slowly to evaporate the water. Cover with a watch glass loosely. From time to time stir with a glass rod so make sure you don't have clumps in the sand trapping water. Don't heat too fast, though, or you'll lose some $SiO_2$ by spattering. When it appears dry, remove the watch glass with tongs, and heat strongly for another 3-5 minutes. Cool and weigh. Record mass. **Heat to constant mass.** Weigh. Record. See special technique below.

8. After you have done all the calculations, dispose of the $SiO_2$ in the trash crocks.

9. The beaker contains all the NaCl from the sample. As a check on the percentages obtained by loss of mass, we will evaporate the water and weigh the NaCl also. Weigh a second evaporating dish. Pour a small amount of the wash water into the dish. Heat to evaporate. As the volume decreases add more of the wash water to the evaporating dish. In the end all of the wash water will have been evaporated so that theoretically only the NaCl remains. **Heat to constant mass.** Weigh.

**Special Technique:** The term "heat to constant mass" will be used frequently throughout this course. It means heat, weigh, record, reheat, weigh again, record. This provides experimental evidence that the sample is dry, fully reacted or through doing whatever it is going to do. If one does not heat to constant mass, then one cannot be sure that the sample is ready for a final analysis. Perhaps there is some residual water in the sample. Any conclusions based on a mass would be invalid. Be sure to record all the masses during the heating process. When performing calculations use the last mass determination. A change in mass of **less than 0.020 g** is desired in this experiment.

# Experiment 3
# A Study of Hydrates

Some ionic compounds crystalize with molecules of water within the crystal lattice in a definite ratio. The ions are in the proper ratio and in addition a certain number of water molecules are loosely bound. These compounds are called **hydrates** and the water is termed the **water of hydration.** The ionic compound is named in the usual manner (See Appendix A) and the number of water molecules is designated with a prefix. Some typical formulas of hydrates and their names are

| | |
|---|---|
| $Na_2CO_3 \cdot 10\ H_2O$ | sodium carbonate decahydrate |
| $MnSO_4 \cdot 3\ H_2O$ | manganese (II) sulfate trihydrate |
| $CrBr_3 \cdot 6\ H_2O$ | chromium (III) bromide hexahydrate |

Upon heating the water molecules are released from the crystal and the **anhydrous compound** remains. For example if sodium carbonate decahydrate is heated carefully, all the water of hydration will be released as water vapor and anhydrous sodium carbonate is isolated. For each mole of sodium carbonate decahydrate that is heated, ten moles of water would be released. By careful weighing of the hydrate, heating to release the water of crystalization, and weighing the product, the percent water can be determined. Then the formula of the hydrate can be written.

In this experiment, two hydrates will be observed. In the qualitative experiment, a liquid will be collected when a hydrate is heated. The liquid will be analyzed as well as its reaction with the anhydrous compound. In the quantitative experiment, the percentage of water in an unknown hydrate will be determined. The post laboratory questions will present data to determine the formula of a hydrate when the identity of the anhydrous compound is known.

**Procedure:**

Part A: A Qualitative Study of a Hydrate

***Work in pairs for Part A. Your TA will assign you a colored hydrate.***

The blue compound is copper (II) sulfate pentahydrate.

The green compound is nickel (II)chloride hexahydrate. Observe the results of the compound not assigned to you from a neighboring group so you may generalize your conclusions.

1. In a 6" test tube place about 2 grams of the hydrate assigned to your group. See the test tube taped to the hood for an approximate amount. It is not necessary to weigh in this step. Gently tap the test tube so that all the compound is in the bottom of the tube. You do not want any colored crystals remaining on the upper part of the tube because they will contaminate the liquid that is being collected.
2. Place another clean 6" test tube in an Erlenmeyer flask as a holder.
3. Position the test tube with the hydrate such that any liquid that might form upon heating will flow into the clean test tube as in the figure. The liquid should not be contaminated with colored crystals.

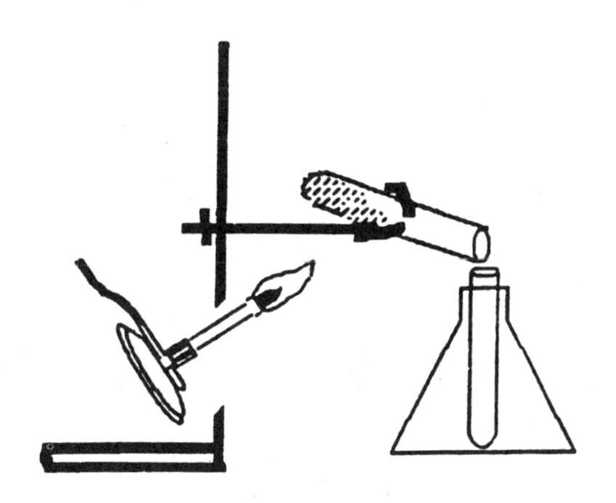

4. Heat the hydrate gently at first. Observe carefully. Continue heating and collect any liquid that might be formed in the clean test tube. Continue heating by moving the burner back and forth with your hand until all of the sample has the same color or no more liquid can be collected. Do not overheat until a black substance is formed. Record your observations.
5. While the samples are cooling prepare a dry piece of cobalt chloride paper by gently warming the paper above a flame. Do not burn the paper, but heat it enough that you are sure it is dry. A color change will be observed. Record the colors of the paper initially and after drying.
6. With a glass stirring rod, place a few drops of the liquid collected from the hydrate onto a portion of the dry cobalt chloride paper. Observe and record. For comparison, add a few drops of distilled water from your wash bottle onto a dry segment of the cobalt chloride paper. What conclusion can you draw?
7. When cool, empty most of the anhydrous compound from the test tube onto a clean watchglass. Divide the solid into two portions. To one portion add a small amount of the liquid from the hydrate with a glass rod. Observe. To the other portion, add a few drops of distilled water from your wash bottle. Observe. What conclusion can you draw?
8. Dispose of the solid in the appropriate copper or nickel waste container.

Part B: Quantitative Determination of the Percentage of Water in a Hydrate

**Do NOT work in partners for this section. Each person will complete an individual experiment. Your TA will assign an unknown to you.**

1. Clean a crucible with nitric acid, rinse thoroughly with distilled water. If a few stains were not removed by this cleaning, they will NOT interfere with the experiment. Continue.
2. Dry the crucible and the lid by heating gently at first and then strongly until the bottom of the crucible has a dull red color. Allow the crucible to cool. (NOTE: If you move a hot crucible, it is likely to break). When the crucible is cool, place it on the wire guaze and carry it to the balance to weigh. Use tongs to place the lid on and off as necessary. Do not touch it with your hands.)
3. Record the mass of the empty crucible and lid. Add about 1 gram of the unknown. Record the exact mass to three decimal places. (Weights between 0.900 g and 1.200 g are appropriate).
4. Heat the hydrate gently at first with the lid ajar to allow water to escape. Do not heat so fast that it spatters. Watch it carefully. As the sample appears more dry, increase the heat. Finally, heat it strongly for about 2 minutes. The bottom of the crucible should be a dull red. Allow it to cool. Weigh. Reheat and weigh to constant mass. (The change should be less than 0.020 grams)

crucible with lid ajar
on clay triangle in ring support

5. Calculate the percentage of water in the compound.

# Experiment 4
# Chemical Equations

A chemist writes an equation to record an observation both qualitatively and quantitatively. The equation contains the correct formulas of all reactants and products and the ratio in which they combine. Equations are the "heart" of chemistry. Two criteria must be met for a correct chemical equation:

a) it must be consistent with the observations
b) it must be consistent with the Law of Conservation of Matter

In recording the observations, each formula must be correct. In addition the physical state should be included if it is not obvious. The term for solid is *(s)*, liquid is *(l)*, and gas is *(g)*. An aqueous solution is noted as *(aq)*. The formulas of the products indicate the chemical change. By balancing the equation the relative amounts of each species is recorded.

To help us write equations, many reactions may be classified as one of the following types:

| | |
|---|---|
| Combination | two or more reactants forming one product |
| Decomposition | one reactant forming two or more products |
| Single Displacement | an element reacts with an ionic compound |
| Double Displacement | an exchange reaction between ions |

However, if we want to focus on the driving forces of reactions, we might also classify reactions accordingly:

| | |
|---|---|
| Precipitation | forming an insouble compound |
| Acid-base (neutralization) | forming a salt and water |
| Gas-forming | forming a water-insoluble gas |
| Oxidation-reduction | transferring electrons |

Some reactions do not fit into the above patterns and some involve a combination of two or more of the above reactions types. Even so, the classification of reactions is useful in learning to write chemical equations.

1. Combination (or Synthesis)
   The combination reaction occurs when two reactants (either elements or compounds) combine totally to form a new compound.

   General Equation: $A + B \longrightarrow AB$

   Examples: $2\,Mg(s) + O_2(g) \longrightarrow 2\,MgO(s)$

   $H_2O(l) + SO_3(g) \longrightarrow H_2SO_4(aq)$

2. Decomposition
   The decomposition reaction occurs when one reactant decomposes into simpler substances, either elements or compounds.

   General Equation: $AB \longrightarrow C + D$

   Examples: $2\,KClO_3(s) \longrightarrow 2\,KCl(s) + 3\,O_2(g)$ gas-forming

   $H_2CO_3(aq) \longrightarrow H_2O(l) + CO_2(g)$ gas-forming

3. Single Displacement

The single displacement reaction occurs when an element and a compound form a different element and a different compound. Both metals and non-metals may undergo single displacement reactions.

General Equation: A + BC $\longrightarrow$ B + AC

Examples: $Sn(s) + Pb(NO_3)_2(aq) \longrightarrow Pb(s) + Sn(NO_3)_2(aq)$

$Cl_2(aq) + 2\ NaI(aq) \longrightarrow 2\ NaCl(aq) + I_2(aq)$

Single displacement reactions may be used to experimentally determine a list with each metal being able to react with a metallic ion of all the metals listed after it. The list is called an **activity series** and is useful in predicting the reactions of many metals. In the case above, tin is more active than lead because when tin metal is placed in a solution of lead ions, such as $Pb(NO_3)_2$, a reaction is observed to produce lead metal and $Sn(NO_3)_2$ Consequently, when the reverse case is considered, Pb metal is placed in a solution of $Sn(NO_3)_2$, *no reaction is observed* because Pb is less active than Sn. By designing experiments carefully the actual number of observations may be minimized by consulting an activity series. For example, if Element X is shown to be more active than tin, it is also more active than lead.

4. Double Displacement

The double displacement reaction occurs when two ionic compounds form two new ionic compounds. There must be a driving force for a double displacement reaction such as formation of a precipitate, evolution of a gas, or the formation of a slightly ionized molecule (eg $H_2O$). Precipitates are solids which settle to the bottom of the test tube. Gases are noted as bubbles. The slightly ionized molecule may be observed if the solution gets hot (or cold) or sometimes when a vivid color change occurs.

General equation: AB + CD $\longrightarrow$ AD + CB

Examples: $Pb(NO_3)_2(aq) + 2\ NaI(aq) \longrightarrow PbI_2(s) + 2\ NaNO_3(aq)$ ***precipitation***

$2\ H_3PO_4(aq) + 3\ Ba(NO_3)_2(aq) \longrightarrow Ba_3(PO_4)_2(s) + 6\ HNO_3(aq)$ ***precipitation***

$HCl(aq) + NaOH(aq) \longrightarrow NaCl(aq) + \mathbf{H_2O(l)}$ **acid/base reaction**

Identification of Products in a Reaction

Much experience is needed to identify the products of a chemical equation. However, a few guidelines will help in this experiment. Some common identification tests are listed in Table 1.

**Table 1: Quick Tests for Common Substances**

|  |  |
|---|---|
| Splint test: | glowing splint ignites in $O_2$<br>burning splint causes $H_2$ to explode<br>burning splint goes out in $CO_2$ |
| Carbonates and Hydrogen carbonates: | will decompose under acidic conditions to form $H_2O$ and $CO_2$ |
| Litmus : | Bases turn red litmus paper blue. Acids turn blue litmus paper red. |

Precipitates can frequently be identified by consulting solubility tables. The short form version in Table 2 is the one most chemists know. However, for specific cases a more detailed chart in Appendix C or the *Handbook of Chemistry and Physics* should be consulted.

**Table 2: Generalized Solubility Rules**

| |
|---|
| All nitrates are soluble. |
| All salts of sodium, potassium, and ammonium are soluble. |
| All chlorides, bromides and iodides are soluble except silver, lead (II), and mercury(I). |
| All sulfates are soluble except barium, strontium, lead (II) and mercury (I). |
| Everything else will be considered *insoluble.* |

## Procedure:

**Work in pairs but be sure to record the data individually. The TA's will stagger the starting point various teams, then work sequentially through the experiment as directed.**

Part A: Zinc and Sulfur

**Use extreme caution:** This reaction is slow to begin but can be very vigorous. Do in groups of 4 in the hood.

Prepare a mixture of powdered zinc and sulfur. Use about one scoop, a sample about the size of a walnut of each. Mix by stirring. Place the mixture on an asbestos pad. Heat with a burner until the sulfur catches fire. Then stop heating and watch. Observe carefully, noting colors at stages of the reaction. Record. Two distinct reactions may be observed.

**Clean-up:** When cool, brush the solid waste into a pile and dispose in the crocks (not in the sink.) Leave the area ready for another group to observe.

Part B: Zinc and Hydrochloric Acid

Place a pea-sized sample of mossy zinc in a test tube. Add about 3 mL of 3 M HCl. Collect a small sample of the gas being evolved by placing an inverted test tube over the reaction. Hold the test tube with the wire test tube holder. Test for combustibility with a lighted splint. This is a classification test for a common gas. What is it? Record your observations and write the balanced equation.

**Clean-up:** Allow the solution to react completely. Dilute with distilled water. If a small piece of solid remains, place it in the crock. If it is all a solution, put it down the sink with excess water.

Part C: Metals in Metallic Ion Solutions

Clean 4 small test tubes. Add 4 drops of the metallic ion solution according to the list below. Then add a small square of the metal. Allow to stand for 15 minutes before drawing any conclusions. Record observations. Look for changes on the surface of the metal. Write "No Reaction" is there appears to be no change. This information is important in developing an activity series. .

tt #1 Zn in $Pb(NO_3)_2$ tt#3 Pb in $Zn(NO_3)_2$
tt #2 Pb in $Cu(NO_3)_2$ tt#4 Cu in $AgNO_3$

Conclusion: A metal which reacts in a metallic ion solution is more "active" than the ion in solution and is placed before the ion in the activity series.

**Waste Disposal:** Decant the solution into the proper waste jar in the hood. Dry off the solid metal before placing it in the crock (not the sink!)

Part D: Neutralization Reactions

Add 5 mL of 6 M HCl to a large test tube. Insert a glass rod and then touch the rod to blue litmus paper. Observe. Put 5 mL of 6 M NaOH in another test tube. Insert a clean glass rod and touch the rod to red litmus paper. Observe. Record the temperature of the acid. Add the NaOH to the HCl. Record the temperature. Retest with litmus. Record. Write the equation for the test tube reaction.

Part E: The Role of the Solvent (water) in Ionic Reactions.

The behavior of $NaHCO_3$ with acid will be observed with and without a solvent (water). "Under acidic conditions" will also be varied with a solid acid, $HC_6H_7O_7$, in Steps 1 and 2 and an acidic solution, 6M HCl, in Step 3. The solution is labeled 6 M HCl which means an aqueous solution of HCl of a 6M concentration. It would be symbolized as HCl(aq).

1. Mix a quantity of solid sodium hydrogen carbonate, $NaHCO_3$, about the size of a lima bean, in 10 mL distilled water. Add 10 mL 6 M HCl and place a lighted splint over the beaker. Observe. Identify the gas based on reaction with the splint (See Table 1).

2. In a 100 mL beaker mix equal amounts of two solids: a quantity of *solid* sodium hydrogen carbonate, $NaHCO_3$, about the size of a lima bean and a similar quantity of a *solid* acid, citric acid, $HC_6H_7O_7$. Stir with a glass rod. Observe. Place a lighted splint over the beaker. Observe and record the changes (if any). This trial has no solvent.

3. Prepare a solution of a lima bean sized quantity of $NaHCO_3$ in 10 mL distilled water in a small beaker. Prepare a similar solution of citric acid in 10 mL water in test tube. Add the citric acid solution to the $NaHCO_3$ solution while you place a lighted splint over the beaker. Observe. Identify the gas based on observations with the splint. Observe and record the changes (if any).

Reaction F: Precipitation Reactions

**Caution:** Do not get silver nitrate on your hands.

Clean three small test tubes.
Reactions 1 and 2: Add 5 drops of silver nitrate, $AgNO_3$, to one test tube and 5 drops of copper(II) nitrate, $Cu(NO_3)_2$, to another test tube. Add 1 drop of 6M HCl to each test tube and observe. Complete the balanced equations if a precipitate is formed.
Reactions 3: Add a *very small amount* of solid NaCl the test tube. Add enough distilled water from your wash bottle to dissolve. Add one drop of $AgNO_3$ and observe.

Part G: A Gas Phase Reaction.

**In the hood** put a bottle of concentrated ammonia ($NH_3$) about 3 inches away from to a bottle of concentrated hydrochloric acid (12 M). Remove the stoppers and observe. Position the two bottles at different distances. Record your results. If possible "collect" some of the product. You can put a large beaker over the two jars and note what forms on the walls of the beaker. Does it dissolve in water?

# Experiment 5
# Properties and Reactions of Solutions Studied by Microscale

In this experiment some physical and chemical properties of some solutions will be studied using microscale techniques. A large number of observations may be made efficiently for comparison and generalization. Refer to the general reaction guidelines from last week's laboratory exercise (Chemical Equations).

## Background Information:

Electricity is conducted by a substance if electrons can flow through it. In some cases (like metals) electrons can travel through the pure solid structure while in other cases the electrons are carried by charged particles (ions) within a solution. Using a battery operated conductivity device assembled as in Figure 1 a number of substances will be tested. The degree of conductivity will be observed by the brightness of a small bulb. Substances which form a conducting solution are called **electrolytes** and those which form a solution which does *not* conduct are **non-electrolytes.**

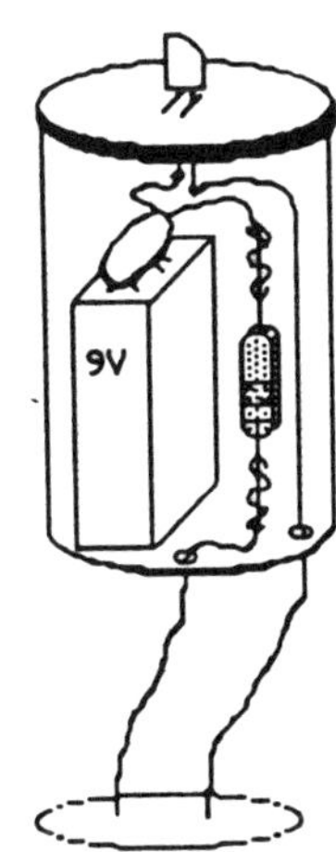


Figure 1. Conductivity Apparatus

Solution conductivity is a measure of the relative number of ions present. If many ions are present in the solution, there is good conductivity. Note that when a precipitate forms, those ions are no longer in solution but rather in the solid structure. However, if spectator ions are present, the solution will still have good conductivity.

The general pattern for a double replacement reaction is

AB + CD --------> AD + CB

Let's consider the reaction of solutions of potassium sulfate and strontium nitrate. After performing the reaction, a white precipitate was observed. The remaining solution conducted an electric current.

Several equations may be written.

Word Equation:

potassium sulfate + strontium nitrate ------> potassium nitrate + strontium sulfate

Symbolic Equation:

$$K_2SO_{4(aq)} + Sr(NO_3)_{2(aq)} \text{--------}> 2\ KNO_{3(aq)} + SrSO_{4(s)}$$

Total Ionic Equation:

$$2\ K^+_{(aq)} + SO_4^{2-}{}_{(aq)} + Sr^{2+}{}_{(aq)} + 2\ NO_3^-{}_{(aq)} \text{---}> 2\ K^+_{(aq)} + 2\ NO_3^-{}_{(aq)} + SrSO_{4(s)}$$

The ions which appear on both sides of the equation may be eliminated from the equation as below. They are called **spectator ions** because they are present but do not participate in the reaction itself. In this case the potassium ions and the nitrate ions are merely in solution.

Net Ionic Equation:

$$Sr^{2+}_{(aq)} \; + \; SO_4^{2-}{}_{(aq)} \; \text{--------->} \; SrSO_{4(s)}$$

Spectator Ions: $K^+$ and $NO_3^-$

The **net ionic equation** is the "bare bones" reaction that is taking place. It is easier to analyze because only the reacting species are written. Chemists often prefer net ionic equations for this reason. Note that sum of the charges on both sides of the equation is the same. The spectator ions are not important in the overall process.

The (aq) in the equation means "in aqueous solution." A chemist realizes that the ions are distributed throughout a large amount of water and that the water molecules are oriented around the ions in a special way. Since $H_2O$ is a polar molecule, the oxygen end of water is near the positive ions it surrounds and the hydrogen end of water is oriented nearer the negative ions. See pictures in your text.

During the pre-lab session, the microscale technique will be presented. Pay close attention to the philosophy and the techniques. Then your laboratory period will proceed smoothly.

**About Microscale:** The chemicals are in 1 mL plastic micropipets or vials. A grid is designed for the experiment and covered with a 96-well plate lid. The reaction is contained within the designated circle on the lid. The results are then recorded in the corresponding space on a larger grid for comparison of several reactions.

Kits: You will be provided with a zip lock bag and four microscale chemical kits.

- a) the Basic Kit
- b) the Conductivity Kit
- c) the Reaction Kit
- d) the pH Kit (no used in this experiment)

The micropipets are labeled and color coded. A key is provided. The concentrations of all solutions are 0.1 M. Keep the pipets in the proper place in the cassette box as indicated in Figure 2. Don't mix them up! Double check the formulas frequently as you work.

Place one drop of solution on the plastic lid in the appropriate space as on Figure 3. Don't touch the solution with the micropipet. Just let it drop onto the lid. Do not get any solution back into the vessel. Keep the chemicals pure!

Observe the reaction. Then proceed to the next one leaving your first reaction on the grid for future comparisons. About 24 reactions will be studied simultaneously.

Leave the drops on the lid for comparison as you work, think, and analyze. Record your observations in the space provided.

To clean up the well-plate, rinse the surface several times with distilled water and dry with towel. Take care not to scratch the surface.

Refill the micropipets (1 mL size) before returning the kits. The technique will be demonstrated in the prelab session. Squeeze the bulb and insert tip into the solution. Release. Turn the bulb to the upright position, expel the remaining air, invert the tip into the solution again to completely fill the vessel. See Figure 4.

**Return the kit so it is ready for reuse.** All pipets must be filled and in order. The well-plates must be clean. Do NOT put the well-plates in the zip-locked bag. Stack them in the area designated by your TA. ***Points will be deducted from your grade if the kit is not returned in proper order.***

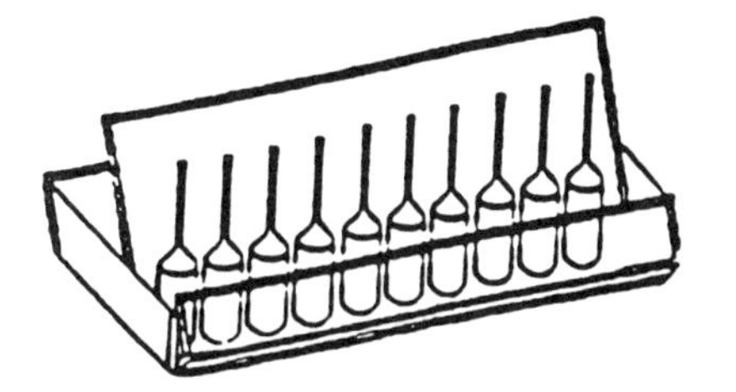

**Figure 2: Microscale Kit**

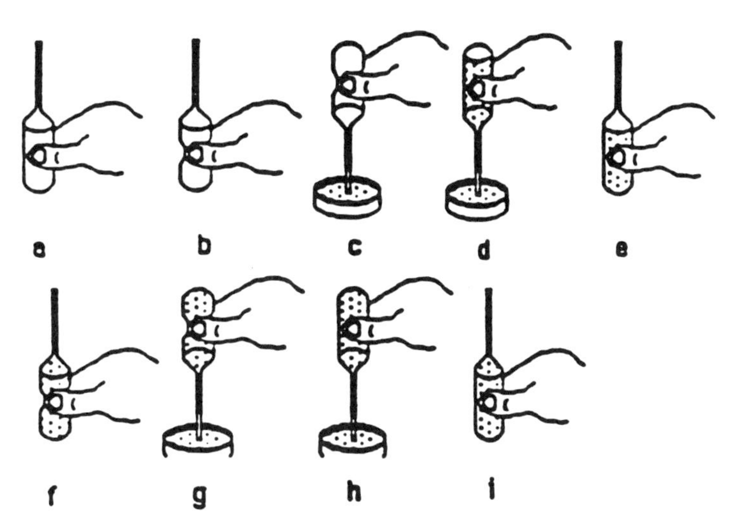


**Figure 4: Technique to Manually Refill Pipettes**

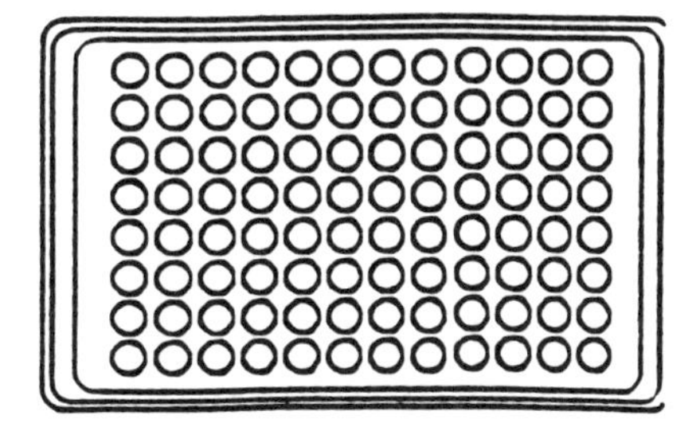

**Figure 3: Well-plate lid**

In this laboratory exercise we will use three cassette boxes with the chemicals as listed. All the concentrations are the same, 0.1 M. Double check the formulas with the names as you are performing the experiments. Take care in writing equations.

1. The Basic Kit -- blue guide sheet

| The Basic Kit | |
|---|---|
| 1) 0.1 M HCl | A) 0.1 M $Pb(NO_3)_2$ |
| 2) 0.1 M NaOH | B) 0.1 M KI |
| 3) 0.1 M $HC_2H_3O_2$ | C) 0.1 M $FeCl_3$ |
| 4) 0.1 M $NH_3$ | D) phenolphthalein |
| 5) 0.1 M $H_2SO_4$ | |
| 6) 0.1 M $NaC_2H_3O_2$ | |

2. The Conductivity Kit -- yellow guide sheet

| The Conductivity Kit | |
|---|---|
| C1) Al | C4) sugar |
| C2) Cu | C5) salt (NaCl) |
| C3) zinc | C6) artificial sweetener |
| wood | **$H_2O$ distilled** |
| graphite(carbon) | tap water |

3. Reaction Kit #1 -- green guide sheet

| **Reaction Kit #1** | | | |
|---|---|---|---|
| #R1 | 0.1 M $Ba(NO_3)_2$ | #R6 | 0.1 M NaI |
| #R2 | 0.1 M $Na_2SO_4$ | #R7 | 0.1 M NaCl |
| #R3 | 0.1 M $Na_2CO_3$ | #R8 | 0.1 M $MgSO_4$ |
| #R4 | 0.1 M $K_2CrO_4$ | #R9 | 0.1 M $AgNO_3$ |
| #R5 | 0.1 M $BaCl_2$ | #R10 | 0.1 M $Cu(NO_3)_2$ |

## Procedure:

Part A: A Study of Conductivity

The conductivity apparatus is a 9-volt battery in a case attached to a small bulb (LED). Two wires (electrodes) protrude from the apparatus. It may be necessary to cup your hand over the bulb to darken the vicinity to see a dim light. If the material being tested conducts electricity, the circuit will be completed and the light glows. The intensity of the light may vary with the solution.

To test a solid substance, place the electrodes on the material and observe. The solid metals may be returned to the appropriate vial after testing. For the crystalline solids, sprinkle a few granules on the plate. Test for conductivity. To prepare a solution, add a few drops of distilled water, mix by gently moving the plate. Test the solution for conductivity. Record results. Clean the plate after completing all of Part A.

Clean the electrodes between trials by rinsing in distilled water.

Sample Data Sheet for Part A (hypothetical student results)

Conductivity Results $H_2O$ wood NaCl(aq) NaCl(s)

**No light**<-----------X--------X-----------X— X--->**Bright light**

Part A Experimental: Compare the conductivity of the items in each set below. Record your results on the brightness scale to rank the items. Label appropriately. If two items appear to have the same conductivity, rank them at the same place on the continuum.

**From Conductivity Kit**

Set 1: copper wire, aluminum wire, zinc, graphite (pencil lead).

Set 2: solids salt (NaCl), sugar ($C_{12}H_{22}O_{11}$), artificial sweetener (Nutrasweet™ or Aspartame, $C_{14}H_{18}N_2O_5$)

Set 3: solutions of salt, sugar, artificial sweetener.
Prepare the solutions by adding a few drops of distilled water to the crystals from Set 2.

Set 4: distilled water, tap water, salt solution, sugar solution
You may use the same solutions prepared in Set 3.

**From Basic Kit**

Set 5: 0.1 M HCl, 0.1 M NaOH, 0.1M $Pb(NO_3)_2$, 0.1 M KI
From Basic Kit

Set 6: 0.1 M HCl, 01. M NaOH, phenolphthalein,
a mixture of 0.1 M HCl, 0.1 M NaOH, and phenolphthalein

Set 7: 0.1 M HCl, 0.1 M $HC_2H_3O_2$, 0.1 M $H_2SO_4$, 0.1 M $NaC_2H_3O_2$

Part B: A Study of the Solubility of Some Sulfates

Cover the circle grid below with the well-plate lid. The top section is a template for Part B. The bottom section is used for Part C. Several empty rows are available in case you need to redo a reaction.

Add one drop of each indicated solution. Do not touch the drops with the pipet. Record your observations on the rectangular grid. If a reaction occurs, identify the products based on observations and the solubility chart. If there is no apparent reaction, you will have learned something about the solubility of the suspected products. For each case where a reaction occurs, write a NET IONIC equation for the process. Identify the spectator ion(s) if any.

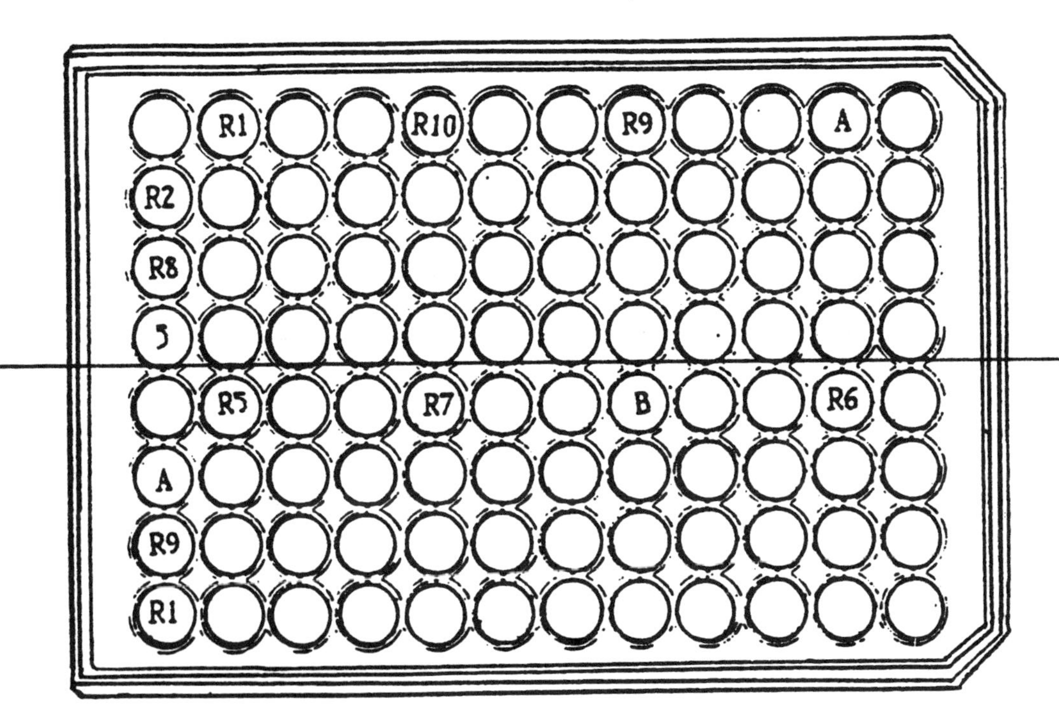


Example: When R1 (barium nitrate) and R2 (sodium sulfate) are mixed, a precipitation reaction might occur. If so, what is the precipitate? There are two possibilities, barium sulfate or sodium nitrate or maybe both of these. Another reaction is needed to truly identify the precipitate. You will be developing some solubility tables <u>based on experiment.</u>

Part C: A Study of the Solubility of Some Chlorides and Iodides

Repeat the procedure from using the solutions as indicated on the grid in the data section.

# Experiment 6
# A Study of Some Oxidation-Reduction Reactions

## Background Information:

The reactivity of an element may be related to its tendency to lose or gain electrons; that is, to be oxidized or reduced. An element that loses electrons undergoes oxidation. An element that gains electrons is reduced. The reagent that causes a loss of electrons by another atom (or ion) is the oxidizing agent while the reagent that causes the gain of electrons is the reducing agent. The oxidizing agent is itself reduced in the process and the reducing agent is oxidized. A careful distinction of these four terms is important in the study of redox reactions.

The reactions of atoms of one element with ions of another element will be explored so that a trend in reactivity may be established experimentally. This is called the activity series and is based on experimental results (not the periodic table or other theoretical scheme)

Any ions present in the solution but not undergoing reaction are termed **spectator ions**. They do not change in the course of the reaction and are omitted from the net ionic equation.

Several *metals* will be placed in solutions of other metallic ions in a manner similar to Experiment 4. This time the pattern of the electron transfer will be the focus of the investigation during the single replacement reaction. If a reaction occurs then the metal will be said to be more active than the metallic ion:

Molecular Equation: A + BC ---> B + AC **if** A is more active than B

Net Ionic Equation: $A^{o} + B^{+}$----->$B^{o} + A^{+}$ Spectator Ion: $C^{-}$

Molecular Equation: D + EF ---> No reaction **if** E is more active than D

By comparing metals two at a time a long activity series may be developed.

Some *non-metals* can be both oxidized and reduced. Sulfur is such a non-metal. The gas $SO_2$ will be generated and dissolved into water. It is unstable in water forming sulfurous acid reversibly.

$$SO_{2\,(g)} + H_2O_{\,(l)} \rightleftharpoons H_2SO_{3\,(aq)}$$

The gas $SO_2$ will be bubbled into several acidic solutions. Sometimes a vivid color change may be observed to indicate a chemical change, other times there is no visible reaction and further tests must be performed. For the purpose of writing equations, consider the aqueous solution of pure $SO_2$. Identify the products from direct or indirect observations. Then balance the equation according to the methods outlined in the lecture and your text.

For example, if $SO_2$ is bubbled through an acidic solution of sodium bromate ($NaBrO_3$) one might postulate the following reaction:

$$H^+_{(aq)} + BrO_3^-{}_{(aq)} + SO_{2\,(g)} \longrightarrow Br^-{}_{(aq)} + SO_4^{2-}{}_{(aq)} \quad \text{(unbalanced)}$$

One of the products must be identified experimentally. The $SO_4^{2-}$ offers a straightforward path for reasoning.

Recall from last week's experiment that aqueous solutions of sodium sulfate and barium chloride can be mixed to produce a white precipitate, $BaSO_4$. By consulting the solubility tables the solid was identified as barium sulfate and the sodium and chloride ions were left in solution The $Na^+$ and $Cl^-$ are **spectator ions**. The net ionic equation for the precipitation is

$$Ba^{2+}{}_{(aq)} + SO_4^{2-}{}_{(aq)} \longrightarrow BaSO_{4\,(s)}$$

This week the same reaction will be used in another manner. The reagent barium chloride ($BaCl_2$) will be added to a solution to test for the presence of $SO_4^{-2}$ in solution. If a white precipitate forms with $BaCl_2$, then $SO_4^{-2}$ was present in the solution. Other oxidized products of $SO_2$ do not form white precipitates.

Some properties of solutions of interest are listed in Table 1:

Table 1: Colors of Aqueous Solutions of Selected Ions and Molecules

| | | | |
|---|---|---|---|
| $I_2$ | yellow/orange | $CrO_4^{2-}$ | yellow |
| $I^-$ | colorless | $Cr^{3+}$ | aqua/green |
| $Cl^-$ | colorless | $SO_4^{2-}$ | colorless |

**Procedure:** ***Work in pairs. Be sure that you see each reaction. Write your own observations and equations.***

Part A: Activity Series of Metals

Calibrate a dropper with water to determine the number of drops in a mL (about 10-12). Then use **half** that number for 0.5 mL volumes in Part A. **Count drops of solutions for volumes.**

Obtain three small pieces of sheet zinc, two of copper and one of lead. Clean the metal with steel wool to expose fresh surfaces. Place six clean 4" test tubes in a rack and add the following reagents:

1: Copper strip and 0.5 mL silver nitrate.
2: Lead strip and 0.5 mL copper (II) nitrate.
3: Zinc strip and 0.5 mL lead (II) nitrate
4: Zinc strip and 0.5 mL magnesium sulfate
5: Copper strip and 0.5 mL dilute sulfuric acid.
6: Zinc strip and 0.5 mL dilute sulfuric acid.

Observe each test tube carefully. Allow to stand at least 15 minutes. Complete the questions and observations on the data sheet **while observing the test tubes.** In other words, don't clean up until you have answered the questions. You may need to make further observations for the activity series.

*Dispose of liquid wastes in proper jars. Place metal pieces in the crocks, NOT THE SINK.*

Part B: Oxidation-Reduction of $SO_2$

1. **Warning:** You may prepare your experiments at your lab bench. But when you are ready to generate the gas, take all materials **to the hood and leave them until you are finished. Keep the waste solutions in the hood also.**

2. In your test tube rack place five clean 6" test tubes.
   Tube 1: Add 5 mL distilled water only.

3. Arrange the other four four 6" test tubes in a rack. Each should contain 5 mL distilled water and 10 drops of dilute HCl. These are *acidic aqueous solutions.* Continue by adding the following to each:
   Tube 2: nothing more. This is a "blank" or "control" test tube.
   Tube 3: 10 drops of 0.1 M iodine (saturated solution bottle)
   Tube 4: 10 drops of 0.1 M potassium chromate solution
   Tube 5: 10 drops of 0.1 M potassium chlorate solution

4. Prepare a sulfur dioxide generator to be used for reactions with the solutions above. Assemble the equipment as in Figure 1 but don't insert the medicine dropper until you are actually ready to generate the gas into the solutions. Place about 3 grams of solid sodium hydrogen sulfite ($NaHSO_3$) in the test tube. Draw as much conc. hydrochloric acid into the medicine dropper as possible. Insert the filled dropper into the hole in the stopper. *Take the assembly to the fume hood.* Clamp the generator to a ring stand. When you are ready to generate the gas, position the medicine dropper and add a few drops of HCl by squeezing the bulb. Refill the dropper with HCl as needed during the experiment.

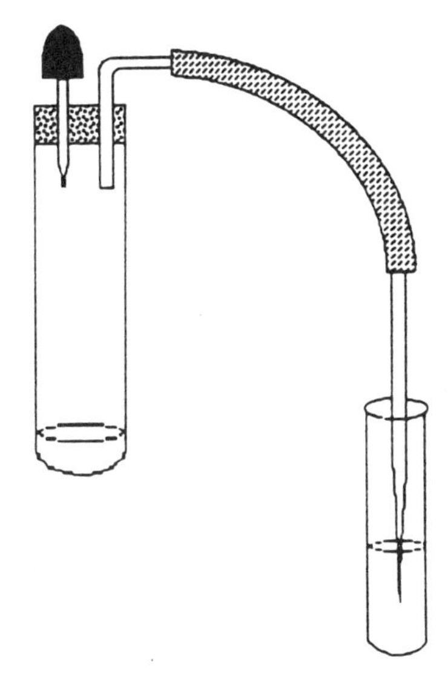

Fig. 1 The $SO_2$ Generator

Bubble $SO_2$ into each of the five test tubes *at a moderate rate* for at least three minutes. Record observations noting how quickly a color change was observed (if any). Sometimes there is visual evidence of redox, sometimes not. It is important to bubble for the indicated time.

Test Tube #1 is merely an aqueous solution of $SO_2$. Test it with litmus paper. Record. Then add 1 mL of $BaCl_2$. Observe. Record your results.

Add 1 mL $BaCl_2$ to Test Tubes 2-5, allow to stand 2-3 minutes and record observations. This is a test for the presence of sulfate ion in solution. It is not part of the redox reaction itself. If you see a precipitate (or a slightly turbid solutions) when barium chloride is added, then sulfate ion is present.

**Leave your test tube rack in the hood. Dispose of the solutions in a waste container in the hood -- not in the sinks.**

4. CAUTION: Hydrogen sulfide is poisonous and offensive. **Please keep the generators in this part in the hood.**

Two generators will be used to study the reaction of $SO_2$ with $H_2S$ in aqueous solution. The two gas generators will be used simultaneously. Check with your TA before beginning this part.

a) Prepare a clean 6" test tube with 5 mL distilled water.

b) Assemble your $SO_2$ generator as before.

c) Use the $H_2S$ generator provided in the hood. Several groups may use the generator for a few minutes as needed. It is a test tube with a one-hole stopper and delivery tube is assembled in the hood with a solid gray material in the test tube. Put 0.5 to 1.0 g of iron (II) sulfide into the test tube generator. Add about 1 mL of **concentrated (12M)** hydrochloric acid and immediately replace the one-hole stopper.

d) Add some HCl to your $SO_2$ generator. Bubble BOTH gases into 5 mL of water for about 2 minutes. Record all observations. Look for a solid substance or a turbid solution.

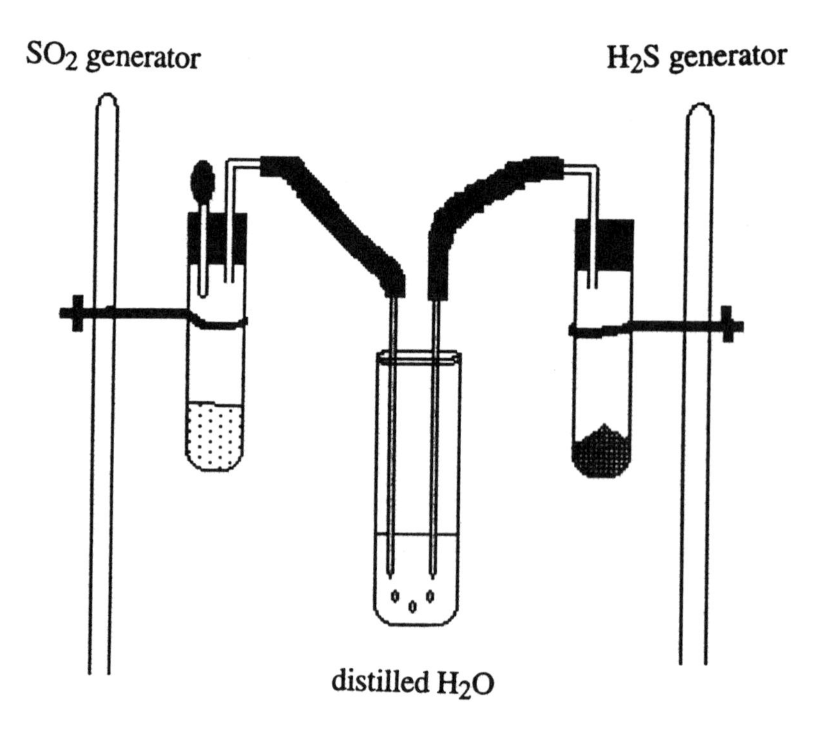

Name ______________________________ Date __________ TA ____________

**Experiment 6: Oxidation/Reduction DATA**

Part A: Activity Series of Some Metals

| Observation | Net Ionic Equation (if any) |
|---|---|
| 1. | |
| 2. | |
| 3. | |
| 4. | |
| 5. | |
| 6. | |

**Conclusion:** Arrange the 5 metals (excluding $H_2$) in a series with the most active first.

Question: $H_2$ was excluded from the activity series because there was insufficient data for proper placement. What additional experiment(s) are needed to properly place $H_2$?

Part B: Oxidation-Reduction of $SO_2$

1. Write the equation for the formation of $SO_2$.

2. Write the equation for the reaction of $SO_2$ with water.(Test Tube #1)

3. Reactions of $SO_2$ under acidic conditions

| Acidic Solution | Initial Observation | Observations after addition of $SO_2$ | Observations after adding $BaCl_2$ |
|---|---|---|---|
| "Blank" no other reagent added | | | |
| $I_2$ | | | |
| $K_2CrO_4$ | | | |
| $KClO_3$ | | | |

4. The $BaCl_2$ is used merely to identify one of the products. It is not part of the redox reaction. Explain why information is gained by observing the reaction with $BaCl_2$.

5. Analyze the reaction of $I_2$ with $SO_2$:

The color change from ______________ to ______________ corresponds to a

______________ of ___________ electron per atom. The oxidation number of
gain/loss how many?

______________ ________________ from ________ to __________.
element increases/decreases

______________ is ________________ and is the ________________ agent.
element oxidized/reduced oxidizing/reducing

Simultaneously ________________(complete the sentence for the other reactant in a similar manner)

6. Analyze the reaction of $K_2CrO_4$ with $SO_2$ using the rationale of the previous reaction. In this case, also identify the spectator ion(s).

7. Analyze the reaction of $KClO_3$ with $SO_2$ using the rationale of the previous reactions. What is different about this reaction? Does this difference in observation change your interpretation of the results? Explain why or why not.

8. Reaction of $SO_2$ with $H_2S$:

Record your observations:

Write the balanced Equation:

Analyze the results of your reaction as in the previous reactions. Note any differences that are apparent in this case.

Name ______________________________ Section ______________ Date ______

**Experiment 6: Redox Reactions**
**Prelaboratory Assignment**

1. In Experiment 4 regarding the single displacement reactions, a student recorded the following observations:
   a) When zinc metal is placed in a lead nitrate solution, a black "fuzz" appears.
   b) When lead metal is placed in a zinc nitrate solution, no reaction is apparent.

   Given: Both zinc and lead form ions with a 2+ charge.

   Write the balanced **net ionic** equation for the reaction of zinc with lead nitrate based on these observations:

   Analyze the net ionic reaction of zinc with lead nitrate by completing the following paragraph:

   The appearance of a black "fuzz" on the surface of the zinc metal corresponds to a ______________ (gain/loss) of ___2___ (how many?) electron per atom. The oxidation number of

   ___zinc___ (element) ________________ (increases/decreases) from ________ to __________.

   ___Zinc___ (element) is _________________ (oxidized/reduced) and is the ________________ (oxidizing/reducing) agent.

   Simultaneously, the oxidation number of __________ (element) ________________ (increases/decreases)

   from ________ to __________.

   ______ (element) is _________________ (oxidized/reduced) and is the ________________ (oxidizing/reducing) agent.

   Which metal zinc or lead is more active? Explain.

over -------->

2. In Experiment 4, the element sulfur was burned in air to give a blue flame. Write a balanced chemical equation for the combustion of sulfur.

Analyze the reaction in a paragraph similar to the Question 1 on the previous page.

3. Determine the oxidation number of sulfur in each of the following:

$H_2S$ ZnS S $Na_2SO_3$ $H_2SO_4$

**Note: In this experiment you will be observing sulfur in several oxidation states.**

# Experiment 7
# Calorimetry

**Background Information:**

Physical and chemical changes are accompanied by changes in energy. The quantity of heat transferred is termed **q**. If heat flows "out of" a system into the surroundings, then $q< 0$ and the process is said to be **exothermic**. On the other hand, if heat flows "in to" the system from the surroundings, then $q > 0$ and the process is termed **endothermic**. In all cases, the law of conservation of energy holds and

$$q_{system} + q_{surroundings} = 0 \qquad \textit{Equation 1}$$

Experimentally we can measure the transfer of heat energy, q, at constant pressure to get a special term the enthalpy change, $\Delta H$. The relationship $q = \Delta H$ applies. The value of q is related to the mass, the specific heat of the substance, and the change in temperature as in Equation 2.

$$q = (\text{mass})(\text{specific heat})(\Delta t) \qquad \textit{Equation 2}$$

A **calorimeter** is a device that is permits one to measure changes in temperature of the medium surrounding a system. Nested styrofoam cups containing a measured amount of water works well. The insulated cup does not permit energy to escape. The physical or chemical change (the system) takes place in the water (the surroundings) and the temperature change ($\Delta t$) of the water is observed. By using the specific heat of water ($4.18\ J/g^0C$ from a handbook) the amount of heat energy gained or lost by the water can be calculated from Equation 2.

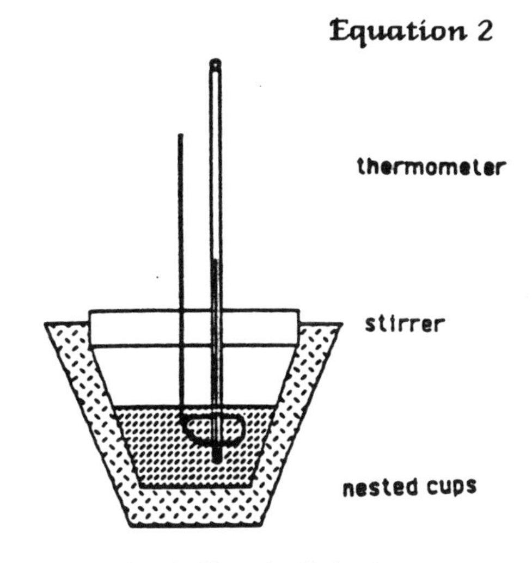


Figure 1: A Simple Calorimeter

Note that the energy change in the system is the algebraic opposite of the energy change in the surroundings, so

heat energy <u>gained</u> by system + heat energy <u>lost</u> by water = 0
or heat energy <u>lost</u> by system + heat energy <u>gained</u> by water = 0

When working at constant pressure, we can say that $q = \Delta H$, the change in enthalpy as diagrammed in Figure 2.

if exothermic: $\Delta H$ is negative
if endothermic: $\Delta H$ is positive

In this exercise three processes, two physical changes and one chemical change, will be studied.

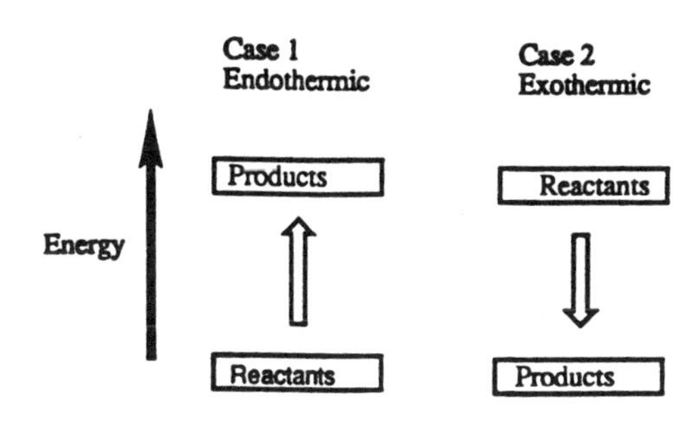


**Figure 2: Enthalpy diagram**

In the first study, a sample of an unknown metal will be heated in a test tube to keep it dry. Then it will be dropped into the calorimeter and the change in temperature of the surroundings (the water) in the calorimeter will be observed. Thus, one can calculate the energy gained by the surroundings and in another step the **specific heat** of the metal. Care must be taken to measure the temperature accurately with a special thermometer.

The second study is designed to measure the total energy change ($\Delta H$) when a salt is dissolved in water. The energy is called the **heat of solution.** The process really consists of two events. The first event, the breaking of the ionic bonds in the crystal lattice, absorbs energy which is called the *lattice energy*. The second event is the surrounding of the ions by water molecules (hydration) which gives off energy. If the *lattice energy is greater than the hydration energy,* the net change is *endothermic*. But *if the hydration energy is greater than the lattice energy*, the process is *exothermic*. You will experimentally determine the heat of solution for an ionic solid.

Thirdly, the reaction of an acid and a base will be considered. The total energy change ($\Delta H$) is called the **heat of reaction.** The energy may be measured similarly in a calorimeter. For example, when sodium hydroxide is neutralized with an acid, heat is given off.

$$\underset{\text{acid}}{HCl} + \underset{\text{base}}{NaOH} \text{ ------> } \underset{\text{salt}}{NaCl} + \underset{\text{water}}{H_2O} + \text{energy}$$

To measure the heat of reaction with proper controls, a system will be set up using a constant amount of sodium hydroxide and varying amounts of different acids. In each case, *the amount of sodium hydroxide will be limiting and the amount of acid in excess.* At the end of each reaction, this will be verified with litmus paper. (In basic solution, red litmus turns blue. In acid solution, blue litmus turns red.) If the total volume of solution in each trial is the same, then the temperature changes will be comparable for each neutralization. For further comparison, a test with additional sodium hydroxide will be conducted.

## Procedure: Work in pairs

Assemble the calorimeter as in Figure 1. It may be necessary to prepare the looped glass rod from soft glass tubing. If so, special instructions will be distributed.

Part A: The Determination of the Specific Heat of a Metal

1. Obtain a metal unknown and a special thermometer from the TA. Use the special thermometer for Part A only.
2. Fill a 500 mL beaker with tap water and heat to boiling. While it is heating, weigh your metal sample in the test tube. Then weigh the empty test tube. Do not handle the metal excessively. Keep it dry. Insert the metal back into the test tube and stopper it LOOSELY.
3. Heat the test tube with the metal in it in the boiling water bath for at least 10 minutes.
4. Measure exactly 40.00 mL of distilled water in a graduated cylinder. Pour carefully into the calorimeter. Reassemble. Record the temperature of the water.
5. Remove the test tube containing the metal from the boiling water bath. Assume the temperature of the metal is now the same as that of the bath, 100°C. Pour the metal quickly into the calorimeter.

6. Stir evenly. Record the maximum temperature reached by the water.

7. Calculate the energy gained by the water using Equation 2 . The energy lost by the metal is assumed to be the same numerically with an opposite sign. Using Equation 1 again, calculate the specific heat of the metal.

8. Repeat if time allows. Return special thermometer to TA.

Part B: The Determination of the Heat of Solution for an Ionic Solid

1. Your TA will assign an ionic solid to you. In a small beaker, accurately weigh out a sample in the range of 5 grams. Record to the nearest 0.001 gram.

2. Assemble a calorimeter with two nested styrofoam cups, a lid, a glass stirrer, and the thermometer from your drawer (<u>not</u> the special one needed for Part A)

3. Measure 50.0 mL distilled water and pour into the calorimeter. Record the temperature as accurately as the thermometer permits. Recall rules for recording significant figures.

4. Add the solid to the water. Replace the cover. **Swirl to dissolve**. (It is imperative that all the solid dissolve to get good results.) Record the change in temperature until no further change is noted.

5. Repeat if time permits.

6. Complete calculations to determine the heat of solution. Watch algebraic signs and significant figures.

Part C: The Determination of the Heat of Reaction for a Neutralization

*The solutions are prepared so the concentrations do not give a clue as to the results. For your information only they are 1.5 M $H_2SO_4$, 3.0 M HCl , 3.0M $HNO_3$, and 5% (by mass) NaOH*

1. Pour the desired volumes of acid and water into the calorimeter. Record the temperature of the mixture.

2. Add the base, cover, stir until the maximum temperature is reached. Record.

3. Test the solution after reaction with litmus.

4. Empty calorimeter. Solutions may be washed down the drain with excess water. Rinse calorimeter with distilled water. Repeat. If results are inconsistent, do another trial to average the results.

**Clean-up:**

Return the cups, lids, and stirrers to the hoods. Do NOT throw them away.
Be sure balance area is properly cleaned. Return special thermometers to TA.

Name ______________________ Date ____________ TA ______

**Experiment 7: Data** *Include proper significant figures and units in all responses.*

Part A: The Determination of the Specific Heat of a Metal

Unknown Number ____________

| | Trial 1 | | Trial 2 |
|---|---|---|---|
| Mass of test tube + metal | ________ | grams | ____________ |
| Mass of empty test tube | ________ | grams | ____________ |
| **Mass of metal** | ________ | **grams** | ____________ |
| Volume of water in calorimeter | ________ | mL | ____________ |
| Density of water | ________ | g/mL | ____________ |
| **Mass of water** | ________ | **grams** | ____________ |
| Temperature of hot metal | 100.0°C | | |
| Final temperature of metal | ______ | | ____________ |
| **Change in temperature of the metal:** | ________ | | ________ |
| Initial temp of calorimeter water | ______ | | ________ |
| Final temp of water | ______ | | ________ |
| **Change in temperature of the water:** (final temp — initial temp) **watch signs!** | ________ | | ________ |
| q water | ____________ | Joules | ____________ Joules (Use Equation 2) |
| qmetal | ____________ | Joules | ____________ Joules |
| Specific Heat of the metal | __________ | J/g°C | __________ J/g°C |

Average Specific Heat __________

Part B: Heat of Solution for an Ionic Solid

| Assigned Solid (formula) |
|---|

Mass of beaker + sample ___________

Mass of empty beaker ___________

**Mass of sample** - - - - - - - - - - -

Volume of water ___________

Density of water ___________

**Mass of water** - - - - - - - - - - -

Initial temperature ___________

Final temperature ___________

**Change in temperature** - - - - - - - - - - -
(final temp — initial temp) ***watch signs!***

Heat transferred to/from water $q_{surroundings}$ = ___________

Heat transferred to/from solid $q_{system}$ =___________

Mass of solid (from above) _________________g

Molar Mass (from formula) _________________g/mole

Moles in sample _________________moles

Heat of solution _________________ J/mol = _______kJ/mol

Conclusion:
The heat of solution for _______________ (formula) is ____________ (sign, value) kJ/mol.

Part C: Heat of Neutralization

| Trial | Acid | Water | Initial Temp. | Base | Max. Temp. | Change in Temp. | After Rxn Acid/Base? |
|---|---|---|---|---|---|---|---|
| 1 | 10 mL HCl | 20 mL | | 10 mL NaOH | | | |
| 2 | 10 mL $HNO_3$ | 20 mL | | 10 mL NaOH | | | |
| 3 | 10 mL $H_2SO_4$ | 20 mL | | 10 mL NaOH | | | |
| 4 | 20 mL HCl | 10 mL | | 10 mL NaOH | | | |
| 5 | 10 mL HCl | 10 mL | | 20 mL NaOH | | | |

1. In each Trial in Part C the total volume of solution was 40 mL. Why is this an important control?

2. The litmus test after reaction provides a clue as to which reagent is limiting and which reagent is in excess. What is the limiting reagent in each case? How do you know?

3. What difference(if any, within experimental limits) is observed when the nature of the acid is changed from HCl to $HNO_3$ to $H_2SO_4$ with the same amount of base? Explain.

4. What happens when the amount of acid is doubled as in Trials 1 and 4? What conclusion may be drawn from your results (within experimental limits)?

5. Compare Trials 1 and 5 in which the amount of base is doubled but the amount of acid remains the same. What conclusion may be drawn?

Name ______________________________ Date ____________ TA _______

**Experiment 7: Calorimetry**
**Prelaboratory Assignment:**

1. When two solutions are mixed, the container "feels" cold. Explain how the energy transfer causes this observation.

2. The Handbook of Chemistry and Physics lists the heat of solution of lithium nitrate, $LiNO_3$ , is – **2.510 kJ/mole.**

$$LiNO_{3(s)} \text{ -------> } Li^{+}{}_{(aq)} + NO_3{}^{-}{}_{(aq)}$$

   a) Sketch the enthalpy diagram for the process. See Fig 2. Does the container "feel" hot or cold? Why?

   b) When 33.5 **grams** of lithium nitrate is dissolved in 80.0 mL water initially at 20.0°C the final temperature of the water may be calculated in several steps. Assume no loss to the surroundings. The density of water is 1.00 g/mL.

      i) Find moles of lithium nitrate in the sample.

      ii) Find the amount of energy transferred to the water based on the handbook heat of solution.

      iii) Calculate the change in temperature of the water from Equation 2.

      iv) Calculate the final temperature of the water.

over ---->

3. A solution containing 0.50 **moles** of potassium hydroxide is mixed with a solution containing 0.25 **moles** of hydrobromic acid. Complete the table below as follows a) write the balanced equation for the reaction with the number of moles available for reaction beneath each reagent. b) how many moles of each reagent *actually react* and how many moles of each product are formed? c) which reagent is limiting? d) which reagent is in excess? e) how could you test this experimentally?

| a)balanced equation | | | | |
|---|---|---|---|---|
| moles available | | | XXXXX | XXXXX |
| b) moles that react / form | | | | |
| moles in excess (if any) | | | XXXXX | XXXXX |

c) LIMITING REAGENT:

d) REAGENT IN EXCESS:

e) experimental test for your conclusions:

# Experiment 8
# A Redox Titration

## Background Information:

The vivid color changes observed in redox reactions can be used quantitatively to determine the amount of substance in a sample. For example, potassium permanganate ($KMnO_4$) is an excellent oxidizing agent and is dark purple in color. It can be used to determine the amount of iron (as $Fe^{2+}$) in an unknown sample because the purple color of the $MnO_4^-$ will disappear when the $MnO_4^-$ is reduced to the colorless $Mn^{2+}$ and simultaneously $Fe^{2+}$ is oxidized to $Fe^{3+}$. The reaction proceeds under acidic conditions according to the equation:

$$\underset{\phantom{purple}}{8\,H^+(aq)} + \underset{\text{purple}}{MnO_4^-(aq)} + 5\,Fe^{2+}(aq) \longrightarrow \underset{\text{colorless}}{Mn^{2+}(aq)} + 5\,Fe^{3+}(aq) + 4\,H_2O(l)$$

The experimental technique for this determination is termed a titration. A weighed amount of solid sample containing $Fe^{2+}$ will be placed in an Erlenmeyer flask. A buret will be filled with a standardized solution of $KMnO_4$ which is slowly added to the flask until the first indication of excess $KMnO_4$. When all the $Fe^{2+}$ from a sample has been oxidized, the purple $MnO_4^-$ remains in solution and is observed as a faint pink color. This marks the **endpoint** of the titration procedure. It is also called the equivalence point.

At the endpoint one can calculate the number of moles of $MnO_4^-$ added from the definition of molarity:

$$\text{Molarity} = \text{moles/volume of solution} \qquad M = \text{mol/liter}$$

and rearranging, $\quad (\text{volume})(\text{Molarity}) = \text{moles}$

Since the moles of $MnO_4^-$ added is related to the moles of $Fe^{2+}$ in solution by the **stoichiometric factor**, one can use the balanced equation to determine the moles of $Fe^{2+}$ in the sample from the equation:

$$(\text{liters of } KMnO_4 \text{ added})(\text{Molarity of } KMnO_4)\left(\frac{5 \text{ moles } Fe^{2+} \text{ consumed}}{1 \text{ moles of } KMnO_4 \text{ added}}\right) = \text{moles } Fe^{2+}$$

**Stoichiometric Factor**

Recall that spectator ions are frequently present in chemical reactions. Solid ionic compounds contain an equal number of positive and negative ions which dissolve in solution. Although the spectator ions are present they do not participate in the redox reaction. In this case the $Fe^{2+}$ will be analyzed in an unknown containing iron(II) ammonium sulfate hexahydrate, $Fe(NH_4)_2(SO_4)_2 \cdot 6\,H_2O$. The ammonium and sulfate ions are spectators as well as the potassium ion from the $KMnO_4$ solution and the negative ions from the acids. Phosphoric acid is added to stabilize the mixture and to prevent any air oxidation during the experiment.

## General Titration Procedure

A buret is an instrument that will deliver a measured volume of a liquid. See Figure 1. It is a long calibrated cylinder with a detachable stopcock assembly. To prepare for a titration the following steps are necessary:

1. Clean the buret with soap and water using a long-handled brush. Rinse with tap water. Reinsert the stopcock assembly. Then rinse again with distilled water. During the rinses rotate the cylinder so that all surfaces are properly cleaned. Finally rinse again with a *small amount* (3-4 mL) of the solution that will be placed in the buret.

2. Fill the buret to a point **above** the 0.00 mL mark. Put a beaker for waste solution under the buret. Open the stopcock to expel any trapped air bubbles. Close the stopcock. **Record the initial volume.** It doesn't have to be exactly 0.00 mL, but you need to record what it is. Values between 0.00 and 5.00 mL are appropriate. Remove the waste container.
3. Begin adding the solution from the buret at a slow to moderate rate while swirling with the other hand. It will take some practice to become proficient at this procedure. Continue to add slowly until some color change is evident. Then proceed very slowly searching for the point when one more drop will cause the desired color change. Frequently it is difficult to discern whether it really is the end point or not. As you near the endpoint, record a **preliminary endpoint,** then add one more drop. If the desired change was observed, **record the endpoint**. If the additional drop was beyond the endpoint, then the preliminary reading was indeed the true endpoint. Once you have gone beyond the endpoint, there is no way to go backwards.

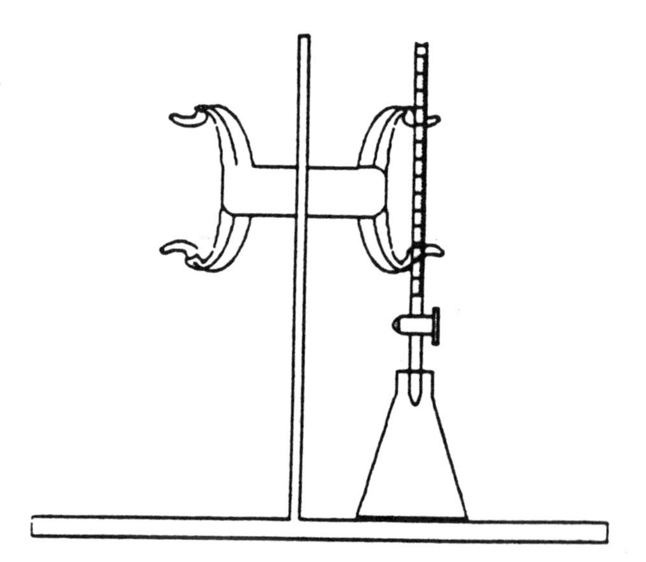

**Fig. 1: Buret in Holder**

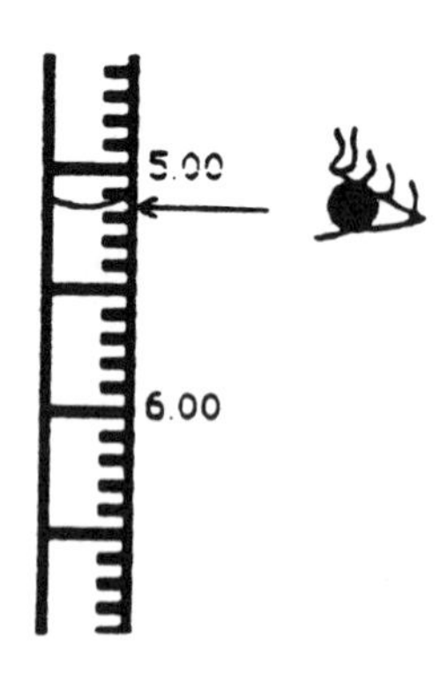


Fig. 2 Read the bottom of the meniscus at eye level. The reading here is 5.15 mL.

4. To record the volume from the buret, position yourself at eye level with the meniscus. See Figure 2. Sometimes placing a piece of white paper behind the buret will improve visibility. Record the bottom of the meniscus to the nearest hundredth.

5. If more samples are to be titrated, add additional reagent to a point near the top, record the initial volume for the next trial, and continue. It is not necessary to clean the buret between samples.

6. At the end of the session, clean the buret with soap and water, Rinse with distilled water. Return.

**Procedure:**

**Part A. Preparation of $KMnO_4$ Solution.**

1. Obtain approximately 25 mL of $KMnO_4$ stock solution from the TA in a clean graduated cylinder.

2. Pour the 25 mL of $KMnO_4$ into a clean Florence flask.

3. Rinse the graduated cylinder with two 10-15 mL portions of distilled water and pour all into the Florence flask. Swirl gently to mix.

4. Add approximately 200 mL of distilled water to the Florence flask. You now have approximately 250 mL of $KMnO_4$ solution. You will now standardize the solution.

**B. Standardization of $KMnO_4$ solution.**

1. Weigh accurately (to three decimal places) two samples of **PURE** $Fe(NH_4)_2(SO_4)_2 \cdot 6\ H_2O$. **Use a mass in the range of 0.8 to 1.0 grams.**

2. Clean the buret as outlined in Step 1 on page 68. Don't forget the last rinse with a small amount (< 5 mL) of the $KMnO_4$ solution. Fill to a point near the top. Record the initial volume accurately on the data sheet.

3. Prepare diluted $H_2SO_4$ by carefully pouring 32 mL of 6 M $H_2SO_4$ into 168 mL of distilled water while stirring in beaker. The solution is now 1 M $H_2SO_4$. **Take special care not to get this solution on your clothing.**

4. Add 50 mL of this 1 M $H_2SO_4$ solution to ONE of the iron samples previously weighed. The sample should dissolve completely. Add 3 mL of 85% $H_3PO_4$ to stabilize the solution. Begin titration immediately. Do NOT let the iron samples stand in acid for a long time as other reactions might occur and alter your results.

5. Titrate by the outlined technique on page 68. Record the volume when a faint pink color persists.

6. Repeat the titration with the other PURE sample. Add the diluted $H_2SO_4$ immediately before the analysis and again add $H_3PO_4$.(Steps 4-5) It is not

necessary to clean the buret again. Just refill with the $KMnO_4$ solution to a point near the top. Don't forget to record the initial volume.

7. Calculate the molarity of your $KMnO_4$ solution. **The molar mass of $Fe(NH_4)_2(SO_4)_2 \cdot 6\ H_2O$ is 392.12 grams**. Find the moles of $Fe(NH_4)_2(SO_4)_2 \cdot 6\ H_2O$ in the weighed sample. Use the stoichiometric factor to find the moles of $KMnO_4$ needed to react with the pure sample. Use the volume you added (in Liters) to find the molarity of the $KMnO_4$. Record the calculated molarity on page 71 and continue.

8. Dispose of the samples by first neutralizing them by adding 20.0 mL of 6 M NaOH. Swirl to mix thoroughly. Then the samples may be discarded in the sink with plenty of excess water. Our filtering beds required a large volume of water to work properly.

## C. Determination of the Unknown Sample.

1. Weigh out accurately (to three significant figs) two samples of the **unknown** samples of $Fe(NH_4)_2(SO_4)_2 \cdot 6\ H_2O$. **Use a mass in the range of 0.8 to 1.0 grams**. (The unknown samples contain an unreactive material and the $Fe(NH_4)_2(SO_4)_2 \cdot 6\ H_2O$. We want to determine the % Fe in the unknown sample. It will be different from the pure $Fe(NH_4)_2(SO_4)_2 \cdot 6\ H_2O$ in Part B.)

2. Titrate as before following Steps 4-6 in the above outlined procedure. Record your results on page 71. Do two titrations with the unknown samples.

3. Discard your solutions as outlined in Part B, step 8.

4. Calculate the % of iron in your unknown sample.

Name ______________________________ TA ______________

**Experiment 8: Redox Titration Data**

Part B: Standardization of a freshly prepared $KMnO_4$ Solution

| | Trial 1 | Trial 2 |
|---|---|---|
| Mass of pure $Fe(NH_4)_2(SO_4)_2 \cdot 6\ H_20$ | | |
| Moles of pure $Fe(NH_4)_2(SO_4)_2 \cdot 6\ H_20$ | | |
| Moles of $Fe^{2+}$ in pure sample | | |
| Moles $MnO_4^-$ required to reach the endpoint (use the stoichiometric factor) | | |
| Initial Buret Reading | | |
| Final buret reading | | |
| Volume of $KMnO_4$ used | mL | mL |
| Volume of $KMnO_4$ for calculation | L | L |
| Molarity of $KMnO_4$ | | |

| Average Molarity of $KMnO_4$ |
|---|
| |

Sample Calculations:

Part C: Determination of the % Iron in an Unknown Sample of $Fe(NH_4)_2(SO_4)_2 \cdot 6\ H_20$

Unknown Number _______________

**Mass of Sample #1** _______________

**Mass of Sample #2** _______________

Molarity of $KMnO_4$ solution _______________ (your results from Part B)

| | Trial 1 | Trial 2 |
|---|---|---|
| Initial Buret Reading | | |
| Final buret reading | | |
| Volume of $KMnO_4$ Used | | |
| Moles $MnO_4^-$ required to reach the endpoint | | |
| Moles $Fe^{2+}$ present | | |
| Mass of Fe in sample | | |
| Percent Fe in sample | | |

| Average % Fe in sample: |
|---|
| |

Sample Calculations:

Name ______________________________ Date ____________ TA _______

**Experiment 8: Redox Titration**
**Prelaboratory Assignment:**

*Show complete calculation set-up. Watch significant figures and units.*

**You will be working with sulfuric acid and potassium permanganate solutions in this experiment. Please wear old clothing in case of minor splashing.**

The balanced equation for this reaction is

$$8\ H^{+}(aq) + MnO_4^{-}(aq) + 5\ Fe^{2+}(aq) \text{ -------> } Mn^{2+}(aq) + 5\ Fe^{3+}(aq) + 4\ H_2O(l)$$

1. Suppose 26.35 mL of 0.0122 M $KMnO_4$ are required to titrate the $Fe^{2+}$ in a 0.917 gram unknown sample. Calculate the number of grams of iron in the sample in the following stepwise fashion. Use scientific notation for small numbers.

   a) calculate the number of moles of $MnO_4^-$ added to reach the endpoint.

   ________________

   b) calculate the number of moles of $Fe^{2+}$ in sample using the stoichiometric factor.

   ________________

   c) calculate the number of grams of $Fe^{2+}$ in the sample

   ________________

   d) calculate the % Fe in the sample.

   ________________

3. What is the % iron in **pure** solid $Fe(NH_4)_2(SO_4)_2 \cdot 6\ H_20$? (Our samples are not pure, they have a nonreactive substance also present. However, we do know the exact % composition of the unknown samples. Your results will be compared with the known values.)

# Experiment 9
# Determination of a Chemical Formula

From a balanced chemical equation the relationships between the reactant and products can be determined. Performing such calculations is called **stoichiometry.**

$$2\ KClO_3\ (s) \xrightarrow{\Delta} 3\ O_2\ (g) + 2\ KCl\ (s)$$

When 2 moles of potassium chlorate are decomposed, 3 moles of oxygen and 2 moles of potassium chloride are formed. Had we measured the amounts of reactants and products carefully, we would have been able to see a molar ratio of 2:3:2. Note these are <u>molar</u> ratios *not grams.*

In this experiment we will analyze an unknown compound. By measuring the amount of product formed we will be able to determine the formula of the unknown and write the balanced equation.

The unknowns will be from the following list. with the metal ion given.

The "sodium unknowns" are either $Na_2CO_3$ or $NaHCO_3$
The "potassium unknowns" are either $K_2CO_3$ or $KHCO_3$
The "magnesium unknowns" are either $MgCO_3$ or $Mg(HCO_3)_2$

The unknown will be treated with hydrochloric acid. When the reaction has subsided, the remaining mixture will be heated. The carbonic acid that formed initially will have decomposed into water and carbon dioxide according to Reaction 1 or Reaction 2 below. The carbon dioxide gas bubbles off and after the water is evaporated the solid salt may be weighed. Although excess hydrochloric acid may escape by heating, care will be taken not to add a great excess.

In the case of sodium carbonate the equation is

$$Na_2CO_3\ (s) + 2\ HCl(aq) \text{------>} 2\ NaCl\ (s) + H_2O(g) + CO_2(g) \qquad \textbf{(Reaction1)}$$

Note that the ratio of **moles** of sodium carbonate to **moles** of sodium chloride is 1:2. All of the sodium ions initially present end up as sodium chloride, but all the the carbonate ions have been decomposed.

Similarly, for the compound sodium hydrogen carbonate the equation is

$$NaHCO_3\ (s) + HCl(aq) \text{------>} NaCl\ (s) + H_2O(g) + CO_2(g) \qquad \textbf{(Reaction 2)}$$

This time the ratio of **moles** of sodium hydrogen carbonate to **moles** of sodium chloride is 1:1 and again all the sodium ions are in the solid sodium chloride and the hydrogen carbonate ions have been decomposed.

By weighing the initial mass of unknown and the product sodium chloride, the unknown formula may be determined.

## Theoretical Calculations:

We can now calculate the theoretical masses of product assuming the "sodium unknown' is one or the other salt. Then when we perform the experiment carefully, our results will match one of the theoretical calculations. We can draw a conclusion as to the original composition of the "sodium unknown" as sodium carbonate or sodium hydrogen carbonate.

For quick reference, the Molar Masses are listed

$Na_2CO_3$ =106.01 $NaHCO_3$ = 84.01 NaCl = 58.45

Suppose a student treated 1.000 g of "sodium unknown" with HCl. How many grams of NaCl would be isolated if the salt is assumed to be sodium carbonate?

Based on Reaction 1 (previous page) we calculate :

$$1.000 \text{ g } Na_2CO_3 \times \frac{(1 \text{ mole})}{(106.01 \text{ g})} \times \frac{(2 \text{ moles NaCl formed})}{(1 \text{ mole } Na_2CO_3 \text{ reacts})} \times \frac{58.45 \text{ g NaCl}}{1 \text{ mole}} = 1.102 \text{ g NaCl}$$

Note a **net increase in mass** when one carbonate ion is replaced by 2 chloride ions.

Suppose a student treated 1.000 g of "sodium unknown" with HCl. How many grams of NaCl would be isolated if the salt is assumed to be sodium hydrogen carbonate?

Based on Reaction 2 (previous page) we calculate:

$$1.000 \text{ g } NaHCO_3 \times \frac{(1 \text{ mole})}{(84.01 \text{ g})} \times \frac{(1 \text{ mole NaCl formed})}{(1 \text{ mole } NaHCO_3 \text{ reacts})} \times \frac{(58.45 \text{ g NaCl})}{(1 \text{ mole})} = 0.6952 \text{ g NaCl}$$

This time when the hydrogen carbonate ion is replaced by the chloride ion there is a **decrease in the net mass** of isolated salt.

Use these calculations to summarize the results in Tables 1 and 2.

Table 1: Summary for the Case of the reaction of **1.000 grams Sodium Carbonate**

| Molar Mass 106.01 | 36.46 | 58.45 | 18.02 | 44.01 |
|---|---|---|---|---|
| $Na_2CO_3$ (s)<br>Molar Ratio 1 | + 2 HCl(aq) ------><br>2 | 2 NaCl (s) +<br>2 | $H_2O$(g) +<br>1 | $CO_2$(g)<br>1 |
| Moles $Na_2CO_3$ 1 per formula unit | | Moles NaCl 2 per formula unit | | |
| Moles $Na^+$ 2 per formula unit | | Moles $Na^+$ 2 per formula unit | | |
| Moles $CO_3^{2-}$ 1 per formula unit | | Moles $Cl^-$ 2 per formula unit | | |
| Grams $Na_2CO_3$ **1.000** | | Grams NaCl 1.102 g | | |

Table 2: Summary of the Case of **1.000 grams Sodium Hydrogen Carbonate**

| Molar Mass 84.01 | 36.46 | 58.45 | 18.02 | 44.01 |
|---|---|---|---|---|
| $NaHCO_3$ (s) | + HCl(aq) ------> | NaCl (s) + | $H_2O$(g) + | $CO_2$(g) |
| **Molar Ratio** **1** | **1** | **1** | **1** | **1** |
| Moles $NaHCO_3$ per formula unit 1 | | Moles NaCl per formula unit 1 | | |
| Moles $Na^+$ per formula unit 1 | | Moles $Na^+$ per formula unit 1 | | |
| Moles $HCO_3^-$ per formula unit 1 | | Moles $Cl^-$ per formula unit 1 | | |
| Grams $NaHCO_3$ **1.000** | | Grams NaCl 0.6952 | | |

For your pre-lab assignment you are asked to do similar calculations for 1.000 grams $MgCO_3$ and 1.000 grams $Mg(HCO_3)_2$. Summarize your results in Table 3 and 4.

Table 3: Summary of the Case of 1.000 grams of Magnesium Carbonate

| Molar Mass 84.31 | 36.46 | 95.20 | 18.02 | 44.01 |
|---|---|---|---|---|
| $MgCO_3$ (s) | + 2 HCl(aq) ------> | $MgCl_2$ (s) + | $H_2O$(g) + | $CO_2$(g |
| Molar Ratio 1 | 2 | 1 | 1 | 1 |
| Moles $MgCO_3$ per formula unit 1 | | Moles $MgCl_2$ per formula unit 1 | | |
| Moles $Mg^{2+}$ per formula unit 1 | | Moles $Mg^{2+}$ per formula unit 1 | | |
| Moles $CO_3^{2-}$ per formula unit 1 | | Moles $Cl^-$ per formula unit 2 | | |
| Grams $MgCO_3$ **1.000** | | Grams $MgCl_2$ 1.129 | | |

Table 4: Summary of the Case of 1.000 grams Magnesium Hydrogen Carbonate

| Molar Mass 146.34 | 36.46 | 95.20 | 18.02 | 44.01 |
|---|---|---|---|---|
| $Mg(HCO_3)_2$ | + 2 HCl(aq) ------> | $MgCl_2$ (s) + | 2 $H_2O$(g) + | 2 $CO_2$(g) |
| Molar Ratio 1 | 2 | 1 | 2 | 2 |
| Moles $Mg(HCO_3)_2$ per formula unit 1 | | Moles $MgCl_2$ per formula unit 1 | | |
| Moles $Mg^{2+}$ per formula unit 1 | | Moles $Mg^{2+}$ per formula unit 1 | | |
| Moles $HCO_3^-$ per formula unit 2 | | Moles $Cl^-$ per formula unit 2 | | |
| Grams $Mg(HCO_3)_2$ **1.000** | | Grams $MgCl_2$ 0.651 | | |

$\frac{\text{the} - \text{exp}}{\text{theo}} \times 100$

**Procedure:** Work individually (not in pairs).

Part A: Decomposition of an unknown sample with HCl.

**Recall procedures for drying a crucible and heating to constant mass from last week's experiment.**

You will do two determinations. Since you have only one burner, plan your work so while one is heating the other is cooling. Your results will be based on the average of the two determinations

1. Obtain an unknown. The metal ion will be identified so your compound is a "sodium unknown," a "potassium unknown," or a "magnesium unknown."
2. Clean a crucible and cover and dry it by heating for 5 minutes. Allow to cool.
3. Weigh the cool crucible and cover. Record the mass.
4. Add approximately 0.3 grams of unknown. Weigh accurately.
5. Put the crucible on the clay triangle with the cover off. Add 16 drops of 6M HCl, one drop at a time. There will be "fizzing" as the $CO_2$ escapes. You do not want any of the product to spatter out of the crucible. All of the solid should have dissolved after 16 drops of HCl was added. If there is still solid, add 6 more drops of HCl. There should be no more "fizzing." You don't want a large excess of HCl because you will have to heat it to get rid of the excess but you must be sure all the unknown reacts. Check with your TA if you are uncertain. This is the point of most error in the experiment.
6. Heat gently. Watch it carefully. Take the burner away from time to time. Do not boil the water as spattering will cause a loss of product.
7. As the water volume lessens, the heat may be increased. When the sample appears dry, increase the heat. Heat strongly for 10 minutes. Cool. Weigh.
8. Heat to a constant mass as you did in the previous experiment. It may take two or three weighings.
9. Dispose of the solid after the calculations are complete by dissolving in water.

Part B: Confirmation of the Reaction Products by Comparison with Knowns.

**You can start this part as time permits, but keep your samples until you have finished one trial completely in Part A.**

1. Obtain pea-sized samples the salts listed below or those corresponding to your unknown. (If you have a "sodium unknown", then $Na_2CO_3$ , $NaHCO_3$, and NaCl.) Dissolve in 10-15 drops distilled water. Add 6 M HCl until the fizzing (if any) subsides. Record observations.

   tt #1 $Na_2CO_3$ tt #2 $NaHCO_3$ tt #3 NaCl

2. Obtain a new sample of the compound in B-1 which did not produce a fizz with HCl. Dissolve it 10-15 drops distilled water. (Do NOT add HCl) Add one drop of silver nitrate. Observe.
3. After you have weighed one of your samples to constant mass, use your scoopula and place a pea-sized quantity of your product (Data Table #5) in each of two test tubes. Add water to dissolve. To one add 6 M HCl. Observe. To the other, add one drop of silver nitrate. Observe. By comparing these results to your results in Steps B-1 and B-2, you will be able to identify your product conclusively.

Your Name________________________________________________ TA ___________
A Calculation Experiment, done outside of class to be turned in for lab credit. It will require 2-3 hours to read and complete.

## Experiment 10
# The Hydrogen Emission Spectrum

### Background Information:

Reference: Kotz and Treichel Chapter 7 pages 324-332.

Atomic emission spectra, also called line spectra, are known for many elements. After a sample is energized the electrons are in an **excited state** of high energy. In time they relax to the **ground state** giving off energy in "packets" called quanta. Sometimes the relaxation is in one large "packet" of light energy while at other times the energy is released in a series of smaller "packets." This observation, based on the fact that lines of specific wavelength are observed, rather than all the possible wavelengths, provides experimental evidence for the theories that support the modern view of the atom. In this experiment, we will use data, as the early chemists did, to verify the fact that the wavelengths of light emitted are related mathematically to small whole numbers for the hydrogen atom. We will be able to calculate any one of the values of $\upsilon$, $n_{initial}$, and $n_{final}$ for the hydrogen atom when the values of the other two and the Rydberg constant are known. This will enable us to sketch the Bohr orbit and energy level diagram for the changes relating to any line in the emission spectrum in the UV, visible, or IR regions of the electromagnetic spectrum.

Equations: $E = h\upsilon$ $\upsilon\lambda = c$ $\Delta E = E_{final} - E_{initial}$

for the hydrogen atom $\Delta E = (-2.178 \times 10^{-18}\ J)\left(1/n_f^2 - 1/n_i^2\right)$

Planck's Constant $= 6.626 \times 10^{-34}$ J · sec Speed of Light $= 2.998 \times 10^{8}$ m s$^{-1}$

$n_f$= final state the energy state to which the electron falls (it doesn't have to be the ground state)

$n_i$= initial state the highest energy state to which the electron was excited

### The Experiment:

1. Label the diagram below with the terms: low energy, low wavelength, low frequency, high energy, high wavelength, high frequency, infrared region, visible region (colors ROYGBIV). ultraviolet region.

ENERGY:

| radio/TV waves | microwaves | | | | x-rays | γ rays |
|---|---|---|---|---|---|---|

Frequency:

Wavelength:

2. A hydrogen spectrum contains lines in the UV, visible and IR regions corresponding to transitions as in the figure below.

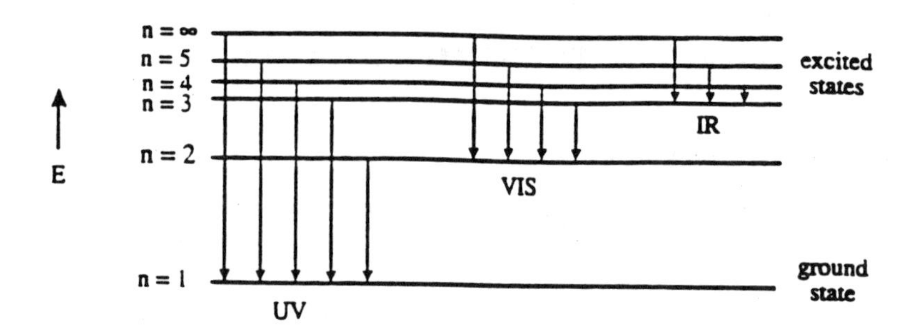


ENERGY LEVEL DIAGRAM FOR HYDROGEN. Changes in energy levels produce lines in the emission spectrum.

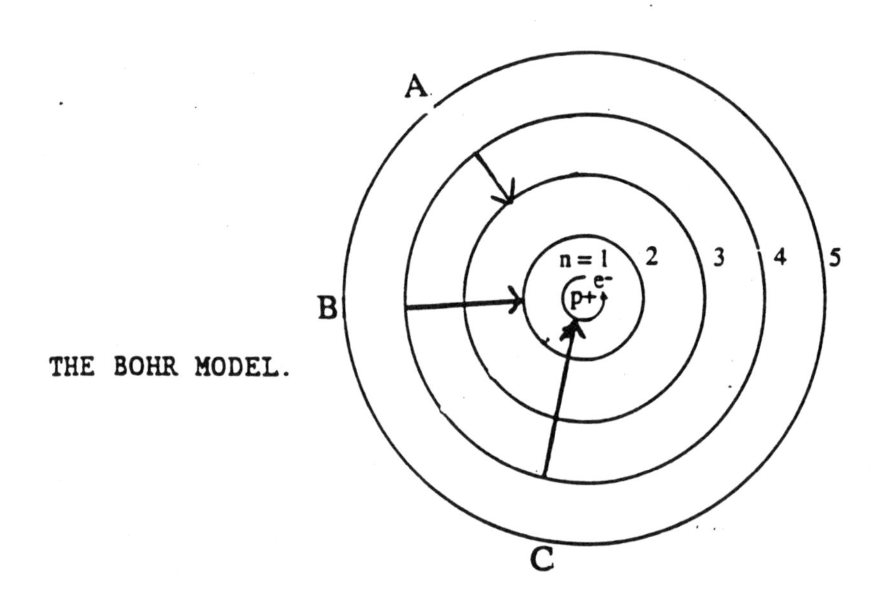


THE BOHR MODEL.

In this experiment we will be concerned with the transitions emitting energy in the visible region. In each case the final energy state is n= 2. These correspond to the transitions originally studied by Bohr, Rydberg and others.

In order to see a transition to the ground state, one must look at lines of the hydrogen spectrum which are in the UV region. Similarly, there are additional lines in the IR region which correspond to transitions from higher orbits to a state where n=3.

## Questions:

a) Match the transitions, A, B, and C on the Bohr model with the corresponding lines of the emission spectrum. Which one is in the visible region? What color is the line? (answer after you have completed the calculations below).

2. (Questions, continued) Determine the color of each line in the visible region for the hydrogen atom. See Kotz, page 326 Figure 7:10. Calculate the frequency of each line with 4 sig figs.

| Wavelength | Color | Frequency ($sec^{-1}$) |
|---|---|---|
| 397.0 nm (not quite in visible region) | | |
| 410.2 nm | | |
| 434.1 nm | | |
| 486.1 nm | | |
| 656.3 nm | | |

Show your work for one line. Do the others similarly on the calculator.

3. Using the frequencies from above and the equation from the first page of this experiment , calculate the transitions for each wavelength of light **given off** by the hydrogen spectrum in the visible region. (Note the sign of light emitted). Express $n_i$ as an integer. Your value should be very close to a whole number.

| Frequency | Wavelength | $\Delta E$ | $n_f$ | $n_i$ |
|---|---|---|---|---|
| | 397.0 nm | | 2 | |
| | 410.2 nm | | 2 | |
| | 434.1 nm | | 2 | |
| | 486.1 nm | | 2 | |
| | 656.3 nm | | 2 | |

Show your work for one transition. Do the others similarly on the calculator.

4. What is the frequency of light emitted when an electron in a hydrogen atom falls from n=5 to n=3? In what region (UV, visible or IR) is this observed?

5. Helium has a spectrum with many more lines. Explain why the situation is much more complex for helium.

## Conclusion:

The green line of the hydrogen spectrum represents the energy emitted by an electron falling from n= __________ to n= ________. After calculating the frequency and using the equation to find the corresponding energy change we see that the energy is in the visible region. Thus the green line represents a final state where n=2. All the other lines of the hydrogen spectrum can be calculated similarly. This gives experimental evidence to the energy transitions theoretically predicted by atomic orbital theories.

# Experiment 11
# A Chemical Family, The Halogens

## Background Information:

Group 7A on the Periodic Chart, the halogens, is an important chemical family. Fluorine, chlorine, bromine, and iodine are the naturally occurring members of the family and the radioactive astatine bears many similarities to the others. Information regarding the physical and chemical properties of the halogens is available in later chapters of your textbook.

The halogens are a very reactive group of *non-metals* which do not occur in an uncombined state in nature, but rather as diatomic molecules or in compounds. Fluorine is extremely reactive and poisonous. It will not be used in this experiment. The other halogens ($Cl_2$, $Br_2$, $I_2$) and their respective halide ions ($Cl^-$, $Br^-$, $I^-$) will be analyzed. The ions exhibit a negative one charge since each has gained an electron so its electron configuration resembles a noble gas. In solution an ionic compound such as sodium chloride contains ions solvated by water as in Figure 1. Throughout this experiment, the positive ions will be spectator ions and will not react.

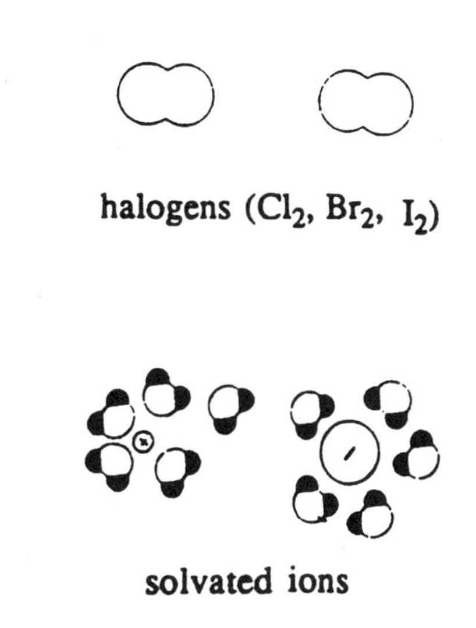


Figure 1. Halogens and Halide Solutions

The halogens undergo some characteristic redox reactions. *They are good oxidizing agents because they are easily reduced to the halide ion by gaining electrons.* This property will be studied qualitatively and quantitatively in the experiment.

The first part of the experiment will be a comparative study of the reactions of halogens ($Cl_2$, $Br_2$, $I_2$), with halide ions ($Cl^-$, $Br^-$, $I^-$). All other ions present will be merely spectators and do not participate in the reaction.

To detect the presence of a halogen, a few drops of toluene will be added. Toluene is immiscible with water and forms a layer on top of the aqueous solution. The halogens are more soluble in toluene than in water and thus will be dissolved preferentially in the toluene layer. A characteristic color in the toluene layer will be used to identify the halogen present. First, some control solutions will be analyzed to determine the color of each halogen in toluene. Then the halogen solutions will be mixed with halide ion solutions in various combinations. The reaction takes place in the aqueous solution, but we must extract the halogen into the toluene layer to make a firm identification. If the halogen is able to take electrons away from the halide ion, a change in color of the toluene layer can be detected to indicate a different halogen and thus a different halide ion (Reactions 1 and 2). However, in some cases there will be no reaction. The experimental data will permit the development of a scheme for the oxidizing ability of the halogens.

$$Cl_{2(aq)} + 2\,I^-_{(aq)} \text{ ---???---> } I_{2\,(aq)} + 2\,Cl^-_{(aq)} \qquad \text{Reaction 1}$$

$$I_{2\,(aq)} + 2\,Cl^-_{(aq)} \text{ ---???---> } Cl_{2\,(aq)} + 2\,I^-_{(aq)} \qquad \text{Reaction 2}$$

The second part of the experiment examines the behavior of the **halide ions** toward the oxidizing agent, potassium nitrite ($NaNO_2$). This reagent is selective and permits one to distinguish between the solutions of the various halide ions. If the ion is oxidized, the corresponding halogen will be observed and identified by comparison with the controls in the previous section. If the ion is not oxidized, no halogen will be detected.

The third part of the experiment involves a useful reaction of the halogen iodine. Iodine will form a deep blue complex with starch (Reaction 4). This reaction is useful to a chemist in two ways. Starch may be added to a solution to test for iodine as in this experiment or iodine may be added to a solution to test for starch as is frequently the test in food analysis.

Starch + $I_2$ -------> deep blue complex (slow) Reaction 3

Vitamin C, also known as ascorbic acid, is a molecule with a number of unusual properties. One property is that it also can react with $I_2$.

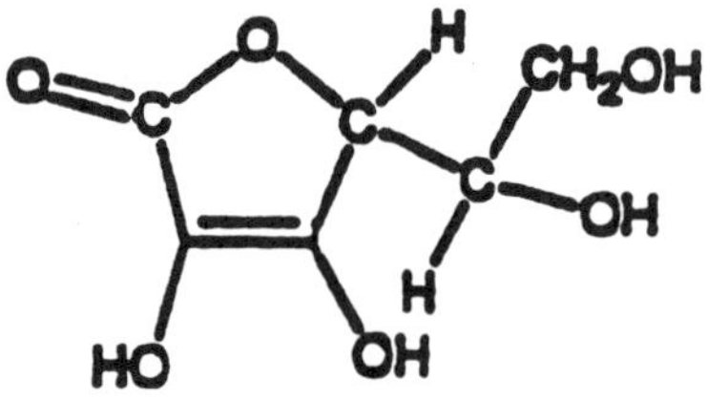

$C_6H_8O_6$

**Short Notation for Ascorbic Acid**

Ascorbic Acid

**Vitamin C**

$C_6H_6O_6$

**Short Notation for Oxidized Product**

The acidic portion of Vitamin C is part of the ring, which undergoes typical acid reactions, not under investigation here. The –OH groups can react in a redox reaction which will be monitored. As the equation below illustrates, the –OH groups can be oxidized fairly easily by iodine. A molecule of iodine ($I_2$) can accept electrons from the Vitamin C and become iodide ions ($I^-$). In the process two $H^+$ are generated into the solution as seen below.

**Reaction 4**

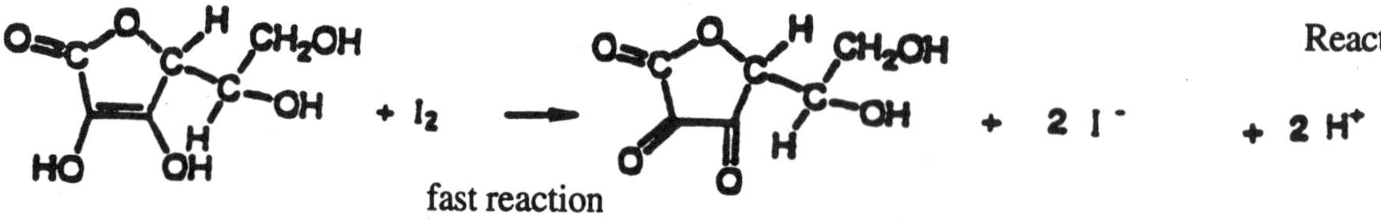

$+ I_2$ —— (fast reaction) $+ 2 I^- + 2 H^+$

**Vitamin C + iodine --------> dehydroascorbic acid + iodide ions + hydrogen ions**

**$C_6H_8O_6$ $C_6H_6O_6$**

When iodine is added to a solution containing Vitamin C the **first** reaction to occur is Reaction 4. The Vitamin C reacts instantly and the iodine disappears. Even though starch is present, no blue color is observed. However, when all the Vitamin C has reacted, there is nothing for the iodine to react with, so it forms the blue complex with the starch solution. By determining how much iodine solution is required to produce a blue color we will know how much Vitamin C was present. **The more drops of iodine that must be added, the more Vitamin C must be present.**

By comparison with a standard of known concentration, we will determine the number of drops of iodine required to generate a blue color. Then we can calculate the amount of Vitamin C present in the sample by comparison.

## Procedure

Part A: Reactivity of the Halogens

1. Prepare the Controls
   In the smallest test tubes available, shake 2 mL of each halogen solution with 2 mL of distilled water. Add 4 drops of toluene. Stopper and shake. Observe the color of the toluene layer **on top** of the aqueous solution. Record on the data sheet. These are the control colors for comparison in other tests. Each halogen ($Cl_2$, $Br_2$, $I_2$) has a distinct color. Save until Parts A and B have been completed. Then dispose in special containers in the hood.

2. Observing the Halogen Molecules

   a. Clean 6 test tubes. Add 2 mL KBr solution to the first test tube. Then add 2 mL chlorine water. Stopper and shake well. The reaction (if any) has now taken place. To detect what happened, we now add 4 drops of toluene, stopper, shake and observe the color of the toluene layer. It will match one of the control colors so you will know if a reaction has taken place.

   b. Repeat the above procedure for the indicated pairs of halide and halogen solutions as on the data sheet. Record the color of the toluene layer on top of the aqueous solution. Do not record the color of the aqueous solution.

   c. Write **balanced net ionic equations** only for those combinations which represent a chemical change. For those that do not, write "no reaction." For example if you added $Cl_2$, and $Cl_2$ was still present after shaking as evidenced by the color of the toluene layer, record "no reaction."

   d. Rank the halogens in an increasing order of oxidizing ability. Recall the activity series from earlier experiments.

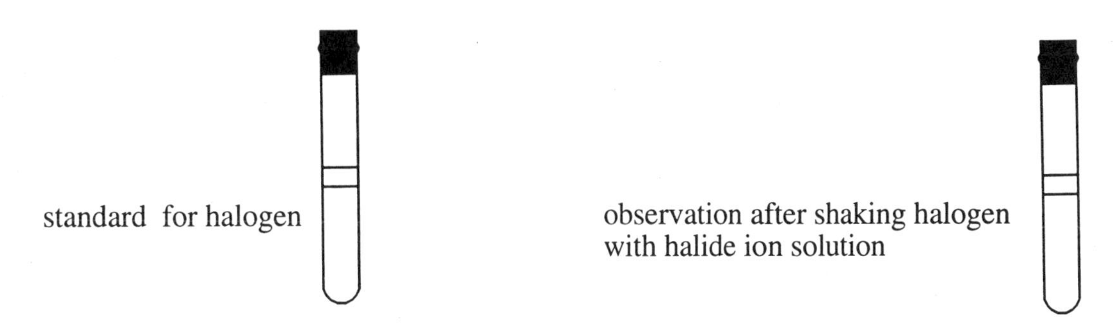

Part B. Selective Oxidation of Halide Ions

1. Place a small quantity (about the size of a pea) of potassium chloride in a test tube. Add 3 mL distilled water and 3 drops of dil HCl. Then add 10 drops of 0.1M sodium nitrite ($NaNO_2$). Stopper and shake. Add 4 drops of toluene, stopper, shake and observe the color of the toluene layer. Recall the control colors of the halogens from Part A1.
2. Repeat with potassium bromide. Repeat with potassium iodide. Record all observations.
3. Repeat with an unknown halide solution. Record the unknown number on the data sheet.

*Dispose of the solutions from Parts A and B in the organic waste container in the hood.*

Part C: Reaction of Iodine with Vitamin C (Ascorbic Acid)

In this microscale experiment you may need to design some of your own procedures. The following is a guide. Depending on the concentrations of Vitamin C in your sample, you may need to adjust the number of drops to obtain a reasonable figure. For example:

if you must add more than 25 drops of iodine to see a color change --- you need to redo the analysis on less sample. (You had too much Vit C present.)

If you saw a color change with fewer than 5 drops of iodine --- you need to redo the analysis with a more sample. (You need more Vit C to analyze.)

1. Obtain a 24-well microplate and micropipets from your TA. Add the following stock solutions which can be identified by color to three different wells.

   a) standard Vitamin C (colorless).
   b) starch (translucent).
   c) iodine (orange).

   Clean your micropipets by rinsing thoroughly in distilled water from a small beaker from your drawer. Assign one clean micropipet to each stock solution.

2. Place the 24-well microplate on a piece of white paper. Organize your experiments so you can record the results appropriately.
3. Place one drop of starch in an empty well. Add one drop of iodine ($I_2$) solution. Observe. This is the reaction when there is NO Vitamin C.
4. Add 4 drops of the standard Vitamin C solution to another well. Add one drop of starch. Count the drops of iodine solution needed to see a color change. Gently shake the well-plate on the paper after each drop is added for good mixing. Record the number of drops it requires for the standard on your Data Sheet. Repeat with 2 drops of standard Vitamin C. Repeat with other quantities until you are certain you know how many drops of iodine are required per drop of Vitamin C standard. See note above about redesigning your experimental procedure.
5. Similarly determine the amount of Vitamin C in **two other solutions** such as fresh orange juice, grapefruit juice, baby juices and brocolli extract. If time permits, you may analyze additional samples.
6. Clean the wells with distilled water. Solutions may be disposed in the sink with excess water. Return the microplates and micropipets to the TA.

Name ________________________________ Date ____________ TA ______

## Experiment 11: The Halogens
## DATA

Part A: Reactivity of the Halogens

Observations (color of the toluene layer)

| | $Cl_2$ | $Br_2$ | $I_2$ |
|---|---|---|---|
| distilled water | | | |

Observations (color of the toluene layer)

| | $Cl_2$ | $Br_2$ | $I_2$ | Net ionic equations (if any) |
|---|---|---|---|---|
| KCl | XXX** | | | |
| KBr | | XXX** | | |
| KI | | | XXX** | |

** no reaction would be expected for this pair

**Conclusion:** Rank the halogens in an increasing order of oxidizing ability. ____________________

Part B: Selective Oxidation of Halide Ions

| | Observations after addition of $NaNO_2$ color of toluene layer | Reaction for Oxidation (if any) |
|---|---|---|
| KCl | | |
| KBr | | |
| KI | | |
| Unknown No. ____ | | |

Part C: Use of Iodine to Determine the Amount of Vitamin C Present in a Sample.

**Concentration of Standard Vitamin C Solution** ________________

Additional determinations

| Solution | Drops of $I_2$ solution | Drops of test solution | Ratio | Amt of Vit C mg/mL |
|---|---|---|---|---|
| Standard Vit C | | | | |
| Sample 1 | | | | |
| Sample 2 | | | | |
| Sample 3 | | | | |
| | | | | |

**Calculation of the Amount of Vitamin C Present in a Sample:**

1. For the standard: Find the ratio of drops of $I_2$ needed to drops of standard used. If it took 12 drops of $I_2$ for 4 drops of Vitamin C, then the ratio is 12/4 = 3.

2. For the unknown: Find the ratio of drops of $I_2$ needed to drops of unknown solution. If it took 24 drops of $I_2$ for 4 drops of unknown, then the ratio is 24/4= 6

Conculsion: The unknown contained 6/3 times more Vitamin C than the standard. If the standard had a concentration of 1 mg/mL, then the unknown had a concentration of 2 mg/mL.

Name ________________________________ Date ___________ TA ______

**Experiment 11: The Halogens**
**Prelaboratory Assignment:**

1. Define the following terms in terms of electrons lost and gained:

   oxidation:

   reduction:

   oxidizing agent:

   reducing agent:

2. Which halogen is most electronegative? How does this property affect its oxidizing ability?

3. Why are the halogens diatomic and the halide ions monoatomic and charged?

3. Consult the solubility tables from Experiment 4 or your textbook. Identify the precipitate when equal volumes of the following solutions are mixed. Assume the concentrations are equal. If there is no formation of a precipitate, write "no reaction."

| | Formula of Precipitate (if any) |
|---|---|
| $NaI$ + $Pb(NO_3)_2$ | ___________________ |
| $NaCl$ + $Mg(NO_3)_2$ | ___________________ |
| $NH_4NO_3$ + $Pb(NO_3)_2$ | ___________________ |
| $NaCl$ + $AgNO_3$ | ___________________ |
| $NaCl$ + $NaI$ | ___________________ |
| $NH_4Cl$ + $Zn(NO_3)_2$ | ___________________ |

## Appendix A:

# General Rules for Naming Chemical Compounds

*You need to use a periodic chart to determine metal/non-metal, charges on ions, and necessity of the Roman Numeral.*

1. Binary Compounds (2 elements)

   a) **two non-metals:** The number of atoms present is indicated by a prefix. The more electronegative element ends in *—ide.*

   Examples: carbon dioxide $CO_2$ diphosphorus pentasulfide $P_2S_5$

   b) **salts,** a metal and a nonmetal. The metal is named first and the non-metal second with an —ide ending.

   Examples: sodium chloride NaCl magnesium phosphide $Mg_3P_2$

2. Ternary Compounds (3 elements)
   Learn the polyatomic groups, formula, charge, and name.

| | | | |
|---|---|---|---|
| $NO_3^-$ | nitrate | $ClO_3^-$ | chlorate |
| $SO_4^{2-}$ | sulfate | | |
| $CO_3^{2-}$ | carbonate | $PO_4^{3-}$ | phosphate |
| $OH^-$ | hydroxide | $NH_4^+$ | ammonium |

3. Acids and Bases

   a) **binary acids** *hydro* —ROOT—*ic acid*

   Examples: HCl(aq) *hydro*chlor*ic* acid $H_2S$(aq) *hydro*sulfur*ic* acid
   The (aq) indicates an aqueous solution

   b) **ternary acids** ROOT *–ic acid*

   Examples: $HNO_3$ nitr*ic* acid $H_2SO_4$ sulfur*ic* acid

   c) **bases** named as metal hydroxide

   Examples: NaOH sodium hydroxide
   $Mg(OH)_2$ magnesium hydroxide

4. Roman Numerals
   If a metal has more than one possible ionic charge, then a Roman numeral is used to specify the positive charge.

   Examples: $Fe(NO_3)_2$ iron (II) nitrate $Bi_2(SO_4)_3$ bismuth(III) sulfate

   $NaClO_3$ sodium chlorate (no Roman numeral because $Na^+$ only)

5. Hydrates
   The number of water molecules loosely attached is indicated by a numerical prefix and the word "hydrate."

   Examples: $CuSO_4 \cdot 5\ H_2O$ copper(II) sulfate pentahydrate

   $MgSO_4 \cdot 7\ H_2O$ magnesium sulfate heptahydrate

6. Mixed Salts
   Salts formed from acids that contain more than one hydrogen atom may replace only one of the hydrogen atoms with a metal. The positive group is named first and the appropriate salt ending.

   Examples: $NaHCO_3$ sodium hydrogen carbonate
   (older method: sodium bicarbonate)

   $Na_2HPO_4$ disodium hydrogen phosphate
   $NaH_2PO_4$ sodium dihydrogen phosphate

Exercise A: Complete the following with formula or name based on the rules above

| Name | Formula | Formula | Name |
|---|---|---|---|
| magnesium bromide | $MgBr_2$ | $H_2Te_{(aq)}$ | Hydrotelleric acid |
| calcium phosphide | $Ca_3P_2$ | $H_3PO_4$ | Phospheric acid |
| ammonium sulfate | $(NH_4)_2SO_4$ | HI(aq) | Hydroiodic acid |
| calcium hydroxide | $Ca(OH)_2$ | $HNO_3$ | nitric acid |
| arsenic (V) bromide | $AsBr_5$ | $OsO_4$ | osmium (VIII) oxide |
| calcium nitrate | $Ca(NO_3)_2$ | ZnS | zinc sulfide |
| potassium chlorate | $KClO_3$ | $Mn(SO_4)_2$ | manganese IV sulfate |
| sodium bromate | $NaBrO_3$ | $SO_3$ | sulfer trioxide |

Exercise B:

| Chemical Name | Formula | Formula | Name |
|---|---|---|---|
| vanadium (II) sulfate | $VSO_4$ | $FeBr_2$ | |
| dinitrogen pentoxide | $N_2O_5$ | $SiO_2$ | |
| iodine | $I_2$ | $Na_3Sb$ | |
| dinitrogen monoxide | $N_2O$ | $(NH_4)_2SO_4$ | |
| lead(IV) nitrate | $Pb(NO_3)_4$ | $H_2SO_4$ | |
| chloric acid | | $N_2O_5$ | |
| hydrobromic acid | | $CuSO_4$ | |
| calcium iodate | | $Ca(OH)_2$ | |
| potassium hydrogen carbonate | | | |
| magnesium hydroxide heptahydrate | | | |
| | | | |

## Appendix B:
## General Rules for Deriving Names and Formulas for the Oxyacids and Salts

a) Learn the names of parent groups

| | Salts | | | Acids | | |
|---|---|---|---|---|---|---|
| | ROOT + ENDING, EXAMPLE | | | ROOT + ENDING, EXAMPLE | | |
| PARENT GROUP | ---ATE | nitrate<br>sodium nitrate | $NO_3^-$<br>$NaNO_3$ | ---IC ACID | nitric acid | $HNO_3$ |

b) Derive other compounds from the parent based on the number of oxygen atoms. There is no change in the charge.

| | Salts | | Acids | |
|---|---|---|---|---|
| | ROOT + ENDING, EXAMPLE | | ROOT + ENDING, EXAMPLE | |
| one more O than parent | per---ate | $ClO_4^-$ $NaClO_4$<br>sodium perchlorate | per --- ic acid | $HClO_4$<br>perchloric acid |
| PARENT GROUP | ---ATE | $ClO_3^-$ $NaClO_3$<br>sodium chlorate | ---IC ACID | $HClO_3$<br>chloric acid |
| one less O than parent | ---ite | $ClO_2^-$ $NaClO_2$<br>sodium chlorite | ---ous acid | $HClO_2$<br>chlorous acid |
| two less O than parent | hypo --- ite | $ClO^-$ $NaClO$<br>sodium hypochlorite | hypo---ous acid | $HClO$<br>hypochlorous acid |

c) Examples:

| Name | Formula | Name | Formula |
|---|---|---|---|
| ammonium phosphite | $(NH_4)_3PO_3$ | calcium chlorite | $Ca(ClO_2)_2$ |
| potassium nitrite | $KNO_2$ | zinc sulfite | $ZnSO_3$ |
| iron (III) hypochlorite | $Fe(ClO)_3$ | lead(II) perchlorate | $Pb(ClO_4)_2$ |
| nitrous acid | $HNO_2$ | hypochlorous acid | $HClO$ |
| carbonic acid | $H_2CO_3$ | sulfurous acid | $H_2SO_3$ |

**Worksheet for Appendix B: Derived formulas**

| Chemical Name | Formula | Formula | Name |
| --- | --- | --- | --- |
| sodium nitrite | | $BaCO_3$ | |
| potassium hypochlorite | | $HClO$ | |
| magnesium sulfite | | $NaNO_2$ | |
| iron(III) nitrite | | $(NH_4)_2SO_3$ | |
| nitric acid | | $HClO_2$ | |
| nitrous acid | | $K_2CO_3$ | |
| sodium hypobromite | | $NaClO$ | |
| calcium periodate | | $Mg(ClO_2)_2$ | |

## Appendix C:
## Using the Generalized Solubility Rules

The general solubility rules are "common knowledge" for a chemist. Although you will be given these rules on exams, they will soon become part of your knowledge base. They are very useful in predicting the course of numerous reactions, particularly the double replacement type and single replacement type.

All nitrates are soluble.
All salts of sodium, potassium, and ammonium are soluble.
All chlorides, bromides and iodides are soluble except silver, mercury (II), and lead(II).
All sulfates are soluble except barium, strontium, lead (II) and mercury (I).

Everything else will be considered insoluble.

Using the solubility rules: When solutions of barium nitrate and sodium sulfate are mixed, a precipitate is observed. What is the formula of the precipitate? Considering this a double replacement reaction, the two possibilities are barium sulfate and sodium nitrate. After consulting the solubility tables we note that sodium nitrate is soluble so the precipitate must be barium sulfate. We write the equation as

$$Ba(NO_3)_2(aq) + Na_2SO_4(aq) \longrightarrow 2\ NaNO_3(aq) + BaSO_4(s)$$

The general solubility rules don't always predict what we need to know, so sometimes a more detailed table must be consulted as below or in your textbook. In addition, the Handbook of Chemistry and Physics in all the laboratories will provide specific data. Some compounds are "slightly soluble" and concentration factors must be considered. A more detailed study of solubilities will be the focus in Chem 152 and Chem 153.

| Class | Solubility in cold water |
|---|---|
| Nitrates | Most nitrates are soluble. |
| Acetates | Most acetates are soluble. |
| Chlorides<br>Bromides<br>Iodides | Most chlorides, bromides, and iodides are soluble except those of Ag, Hg(I) and Pb(II). |
| Carbonates<br>Phosphates | Most carbonates and phosphates are insoluble except those of Na, K, and $NH_4^+$<br>Many hydrogen carbonates(bicarbonates) and acid phosphates are soluble. |
| Sulfates | Most sulfates are soluble except those of Ba, Sr, and Pb; Ca and Ag sulfates are slightly soluble. |
| Hydroxides | Most hydroxides are insoluble except those of the alkali metals and $NH_3$. $Ba(OH)_2$ and $Ca(OH)_2$ are slightly soluble. |
| Sulfides | Most sulfides are insoluble except those of the alkali metal, ammonium, and the alkaline earth metals (Ca, Mg, Ba). |

# NOTES